Giorgio Moresi

Magnetic resonance force microscopy

Giorgio Moresi

Magnetic resonance force microscopy

Interaction forces and channels of energy dissipations

Südwestdeutscher Verlag für Hochschulschriften

Impressum/Imprint (nur für Deutschland/only for Germany)
Bibliografische Information der Deutschen Nationalbibliothek: Die Deutsche Nationalbibliothek verzeichnet diese Publikation in der Deutschen Nationalbibliografie; detaillierte bibliografische Daten sind im Internet über http://dnb.d-nb.de abrufbar.
Alle in diesem Buch genannten Marken und Produktnamen unterliegen warenzeichen-, marken- oder patentrechtlichem Schutz bzw. sind Warenzeichen oder eingetragene Warenzeichen der jeweiligen Inhaber. Die Wiedergabe von Marken, Produktnamen, Gebrauchsnamen, Handelsnamen, Warenbezeichnungen u.s.w. in diesem Werk berechtigt auch ohne besondere Kennzeichnung nicht zu der Annahme, dass solche Namen im Sinne der Warenzeichen- und Markenschutzgesetzgebung als frei zu betrachten wären und daher von jedermann benutzt werden dürften.

Coverbild: www.ingimage.com

Verlag: Südwestdeutscher Verlag für Hochschulschriften GmbH & Co. KG
Heinrich-Böcking-Str. 6-8, 66121 Saarbrücken, Deutschland
Telefon +49 681 37 20 271-1, Telefax +49 681 37 20 271-0
Email: info@svh-verlag.de

Approved by: University of Basel, Faculty of Science, Diss., 2005

Herstellung in Deutschland (siehe letzte Seite)
ISBN: 978-3-8381-2502-2

Imprint (only for USA, GB)
Bibliographic information published by the Deutsche Nationalbibliothek: The Deutsche Nationalbibliothek lists this publication in the Deutsche Nationalbibliografie; detailed bibliographic data are available in the Internet at http://dnb.d-nb.de.
Any brand names and product names mentioned in this book are subject to trademark, brand or patent protection and are trademarks or registered trademarks of their respective holders. The use of brand names, product names, common names, trade names, product descriptions etc. even without a particular marking in this works is in no way to be construed to mean that such names may be regarded as unrestricted in respect of trademark and brand protection legislation and could thus be used by anyone.

Cover image: www.ingimage.com

Publisher: Südwestdeutscher Verlag für Hochschulschriften GmbH & Co. KG
Heinrich-Böcking-Str. 6-8, 66121 Saarbrücken, Germany
Phone +49 681 37 20 271-1, Fax +49 681 37 20 271-0
Email: info@svh-verlag.de

Printed in the U.S.A.
Printed in the U.K. by (see last page)
ISBN: 978-3-8381-2502-2

Copyright © 2012 by the author and Südwestdeutscher Verlag für Hochschulschriften GmbH & Co. KG and licensors
All rights reserved. Saarbrücken 2012

Magnetic resonance force microscopy: Interaction forces and channels of energy dissipation

Inauguraldissertation
Zur
Erlangung der Würde eines Doktors der Philosophie
vorgelegt der
Philosophisch-Naturwissenschaftlichen Fakultät
der Universität Basel
von

Giorgio Moresi
aus Bellinzona (Tessin)

Basel, 2005

Genehmigt von der Philosophisch-Naturwissenschaftlichen Fakultät
auf Antrag der Herren Professoren:

Prof. Dr. E. Meyer
Prof. Dr. H.J. Güntherodt

Basel, den 25 Januar 2005   Prof. Dr. H. J. Wirz, Dekan

to Stefania

Contents

List of Figures VI
List of Tables VII
List of Symbols VIII

Introduction   **1**
   Outline of the thesis 3

1 Magnetic resonance force microscope   **4**
   1.1   Beam deflection 4
   1.2   Magnetic resonance force microscope detection 6

2 Design and construction   **8**
   2.1   Motivations 8
   2.2   The Ultra High Vacuum (UHV) machine 9
       2.2.1   The measurement region 10
       2.2.2   The cooling process 16
       2.2.3   Sample rod and mechanical noise filtering 21

3 Cantilevers: Thermodynamic equilibrium and quality factor   **26**
   3.1   Theory: Harmonic mechanical oscillator in thermodynamic equilibrium   26
       3.1.1   The Gaussian distribution 27
       3.1.2   The autocorrelation 29

4 Damping losses of cantilevers   **33**
   4.1   Determination of the tip mass 33
   4.2   Damping losses of cantilevers in an external magnetic field 36
       4.2.1   Theory: Interaction between magnetic tip and external magnetic field 36
       4.2.2   Frequency shift and damping factor 38
   4.3   The energy losses and the damping factor 40
       4.3.1   Losses due to the oscillating magnetic field 41
       4.3.2   Tip-field interactions 42
       4.3.3   Magnetic interaction losses 45

5 Experiments: Tip-field interactions   **46**
   5.1   Tip materials and setup 46
   5.2   The frequency shift as a function of the magnetic field 49
   5.3   The quality factor measurement as a function of the magnetic field . 50

	5.4 The force sensitivity as a function of the magnetic field	53
6	**Tip-sample interactions and damping losses**	**56**
	6.1 Tip-sample interactions	56
	6.1.1 Frequency as a function of the distance: theory and measurement	56
	6.1.2 The axial force and the frequency shift	58
	6.1.3 The electrostatic force	61
	6.1.4 The diamagnetic force	63
	6.2 Measurement: tip-sample interactions	67
7	**Signal to noise ratio in continuous wave magnetic resonance force microscopy**	**73**
	7.1 Theory: Signal to Noise Ratio in electron spin resonance	74
	7.1.1 Inductive coupled continuous wave Electron Spin Resonance Signal to Noise Ratio.	75
	7.2 Magnetic resonance force microscopy and signal to noise ratio	78
8	**Electron spin resonance at room temperature**	**80**
	8.1 CW-ESR Inductive Experiments	80
	8.2 CW magnetic resonance force microscopy	83
	8.3 Signal to noise ratio in MRFM and CW-ESR	87
	8.4 Discussion and conclusions	88
9	**General conclusion**	**89**
	9.1 Designing and building the MRFM machine	89
	9.2 Improving the sensitivity	89
	9.3 Reduction damping losses	90
	9.4 Magnetic resonance force detection	90
Bibliography		**92**
Acknowledgements		**96**
A	**Annexes**	**97**
	A.1 Beam Flexure: Including axial force effect	97
	A.2 Design of RF coax cable	99
	A.3 Cryogenic System.	100
Curriculum vitae		**103**

List of Figures

1.1	Magnetic resonance force microscopy setup.	5
1.2	Magnetic resonance force microscopy sensor: Sketch and picture . .	7
2.1	Machine photograph	10
2.2	Overview of the magnetic resonance force microscope	11
2.3	The middle chamber	12
2.4	Middle chamber block scheme	13
2.5	The top chamber	13
2.6	The sample rod	15
2.7	Side view of the cryostat	16
2.8	Thermal conductivity	17
2.9	Resistivity	18
2.10	Power transfer	19
2.11	Radiation power transfer as a function of the temperature	21
2.12	Microscope and baffles temperature	22
2.13	The eddy current damping system	23
2.14	The eddy current damper decay	24
2.15	The spectrum of the excited eddy current damping	25
2.16	Transfer function of the eddy current damping	25
3.1	Gaussian probability distribution	27
3.2	Probability distribution	28
3.3	Amplitude oscillations	29
3.4	Thermal noise	30
3.5	The Autocorrelation	30
3.6	The autocorrelation and the Hilbert Transformation	32
4.1	Tipped cantilevers	34
4.2	Nanosensor eigenfrequency vs mass	35
4.3	IBM eigenfrequency vs mass	35
4.4	Tipped cantilever in a magnetic field	37
4.5	Cantilevers and dissipations	40
4.6	Energy dissipation processes	41
4.7	Tipped cantilevers in a static field	43
4.8	Hysteresis loops	43
4.9	Hysteresis loop of FeCo	44
5.1	Tip-field interaction setup	49

5.2	Frequency shift vs magnetic field	50
5.3	Normalized Q factor vs magnetic field	51
5.4	Magnetic friction as a function of the magnetic field	52
5.5	Magnetic friction vs frequency oscillation	53
5.6	Sensitivity as a function of the magnetic field and Q factor	54
5.7	Sensitivity as a function of the magnetic field and frequency shift	54
6.1	IBM cantilever with magnetic tip	57
6.2	Force gradient and force interaction	58
6.3	Axial tensile force	59
6.4	Axial compressive force	60
6.5	Axial tensile and compressive force	60
6.6	Electrostatic force as a function of the distance	62
6.7	Frequency shift as a function of the potential	62
6.8	Magnetic field and gradient at 100 nm	64
6.9	Magnetic field and gradient at 2500 nm	64
6.10	Magnetic field and gradient at 5500 nm	65
6.11	Magnetic dipole field	66
6.12	Magnetic dipole gradient	66
6.13	Diamagnetic force as a function of the distance	67
6.14	Frequency shift as a function of the distance in horizontal approach	68
6.15	Frequency as a function of the distance	69
6.16	Q factor as a function of the distance	70
6.17	Friction as a function of the distance	70
6.18	Power spectral density as a function of the distance	71
6.19	Charge spectral density as a function of the distance	72
7.1	Continuous wave electron spin resonance experiment	76
8.1	Inductive ESR scheme	81
8.2	Absoption line of DPPH	82
8.3	Photograph of the MRFM microscope's head	83
8.4	Scheme of the MRFM detection	84
8.5	Photograph of sample	85
8.6	MRFM signal	86
8.7	SNR of MRFM and IESR	87
A.1	Coax cable power transmission	99

List of Tables

2.1	Power losses due to thermal conduction	19
4.1	Magnetic properties	38
5.1	Cantilevers and tip characteristics	47
5.2	Tipped cantilevers	48
7.1	Inductive sensitivity	77
7.2	Force sensitivity	79
8.1	Sample measured by inductive methods	82
8.2	DPPH mounted on Nanosensors cantilevers	85

List of Symbols

Absorption line	$\Delta\nu[\text{Hz}]$
Alternating magnetic field	$H_{ac}[\text{T}]$
Alternating RF magnetic field	$B_1[T]$
Amplitude oscillation peak	$x_{pk}[m]$
Angular frequency	$\omega[\text{Hz}]$
Anisotropy constant	$K_1[MJ/m^3]$
Anisotropy energy	$E_{Anisotropy}(\theta_m)[J]$
Avogadro number	$N_A = 6.022 \times 10^{-22}[1/mol]$
Axial force	$\mathbf{N}[N]$
Band pass frequency	$\Delta F[Hz]$
Cantilever bending angle	$\theta[\,]$
Cantilever length	$l[m]$
Cantilever thickness	$t[m]$
Capacitance	$C[F]$
Charge fluctuation	$\delta_e[C/Hz]$
Coil resistance	$R_c[\Omega]$
Component of magnetization in rotating frame	$M_{rx}[A/m]$
Component of magnetization in rotating frame	$M_{rz}[A/m]$
Component of magnetization in rotating frame	$M_z[A/m]$
Cross section area	$A[m^2]$
Cross section conductor	$S[m^2]$
Current	$i[A]$
Demagnetization perpendicular factor	$D_\perp[\,]$
Demagnetization parallel factor	$D_{//}[\,]$
Density	$\rho[kg/m^3]$
Diameter conductor	$\Phi[m]$
Electric permeability	$\varepsilon_0 = 8.8541 \times 10^{-12}[As/(Vm)]$
Elementary charge	$e = 1.602 \times 10^{-19}[1/C]$
Emissivity	$e[\,]$
Energy	$\text{E}[J]$
Force	$F[N]$
Frequency sweep	$F_{sweep}[Hz]$
Friction	$\Gamma[kg/s]$

Gas density	$\rho_{gas}[kg/m^3]$
Gravitational accelaration	$g = 9.81[m/s^2]$
Heat power losses	$P[Watt]$
Heat transfer	$\dot{Q}[J/s]$
Imaginary magnetic permeability	$\mu_i[VS/(Am)]$
Impedance	$Z[\Omega]$
Latent heat	$L_{heat}[J/kg]$
Length	$l[m]$
Liquid density	$\rho_{liquid}[kg/m^3]$
Longitudinal relaxation time	$T_1[s]$
Magnetic charge density	$\sigma[A/m^3]$
Magnetic field	$B[T]$
Magnetic field for unity current	$B_{ud}[T/A]$
Magnetic moment	$m[A/m]$
Magnetic permeability	$\mu_0 = 4\pi \times 10^{-7}[VS/(Am)]$
Magnetic sphere radius	$a[m]$
Magnetic torque	$\vec{\tau}(\theta)[Nm]$
Magnetization	$M_s[A/m]$
Magnetization rotation angle	$\theta_m[]$
Mass	$m[kg]$
Moment of area	$I[m^4]$
Noise	$n(t)[V]$
Normal unitary vector	$\vec{n}[]$
Polarizing static magnetic field	$B_0[T]$
Potential spring energy	$E(\theta)[J]$
Power spectral electric field	$S_E[V^2/m^2]$
Quality factor	$Q[]$
Resistivity	$\rho[\Omega m]$
RMS noise	$n_{rms}[V]$
Signal	$s(t)[V]$
Signal to noise ratio	$SNR[]$
Skin depth	$\delta[m]$
Spring constant	$k_F[N/m]$
Spring constant	$k_0[N/m]$
Stefan Boltzmann constant	$\sigma = 5.68 \times 10^{-8}$ [Watts/(m^2 K^4)]
Susceptibility	$\chi_0[]$
Sweep rate	$R_{sweep}[Hz]$
Temperature	$T[k]$
Thermal conductivity	$k[Watt/(mk)]$
Thermal resistance	$R[k/Watt]$
Time delay	$\tau[s]$
Time period	$T_p[s]$
Total friction	$\Gamma_0[kg/s]$
Total potential energy	$E(\theta, \theta_m)[J]$
Transverse relaxation time	$T_2[s]$

Volume	$V[m^3]$
Young's modulus	$E[N/mm^2]$
Zeeman energy	$E_m(\theta, \theta_m)[J]$

Introduction

Today, smaller and smaller electron and nuclear magnetic resonance structures are extensively studied both from an applied and from a fundamental point of view. The powerful tool of magnetic resonance imaging (MRI) has demonstrated that it is possible to visualize subsurface three dimensional structures with micrometer resolution [1] containing 10^{12} nuclear spins; nuclear magnetic resonance (NMR) spectroscopy has the capacity to determine the three dimensional structure of biological macromolecules [2]. Owing to the larger gyromagnetic ratio of electrons as compared to paramagnetic nuclei, electron spin resonance (ESR) has pushed detection sensitivity to 10^7 spins [3]. Finally, a single electron spin [4] has been detected by magnetic resonance force microscopy (MRFM), employing a device which combines two sensing technologies, namely magnetic resonance imaging (MRI) and atomic force microscopy (AFM). The ultimate goal of MRFM is to map the interior of a material sample, such as a complicated semiconductor structure or a bio-molecule, at atomic scale resolution.

The idea of introducing MRFM to improve the detection sensitivity down to a single spin and thus to resolve atoms of proteins [5],[6] was originally proposed in 1992. Ten years later, Rugar and co-workers reported the detection of a single electron spin resonance in a silica substrate with paramagnetic defects, using a magnetic resonance force microscope [4] with a lateral resolution of 25 nm in one dimension. To achieve this single spin detection, the magnetic resonance force microscopy uses a soft cantilever with a tiny hard magnetic tip material. The inhomogeneity field $\mathbf{B}_{inhom}$ generated from the magnetic tip is superimposed with the homogenous magnetic field $\mathbf{B}_0$ which polarizes the sample. For a radio frequency field the resonance condition is fulfilled in the region where $\omega_1 = \gamma(\mathbf{B}_0 + \mathbf{B}_{grad})$ and where γ is the gyromagnetic ratio of electron or proton. Consequently, the next foreseeable step is to detect a single nuclear spin. In fact, the correspondence between ESR and NMR is very close, and much of the basic theory of ESR is directly applicable to NMR. ESR requires an unpaired electron whereas NMR requires an unpaired nuclear spin for detection. Furthermore, an external static magnetic field is necessary in both ESR and NMR detection. The major difference between the two techniques is due to the gyromagnetic ratio of the proton and electron. ESR entails the higher electron gyromagnetic ratio, as compared to the nuclear gyromagnetic ratio involved in NMR and the sensitivity of EPR is correspondingly higher (approximately a factor of 1000).

The force generated by a single spin is in the attonewton range. Thus, non commercial, soft single crystalline silicon cantilevers with a high quality factor and minimized spring constants have to be used for detecting a single spin. Measurements are performed at liquid helium temperature where thermal noise is reduced by a factor of 10. The UHV condition makes for a very stable environment reducing the oxidation of the sample and of the cantilever. In our low temperature force microscope force sensitivities on the order of 10^{-18} N/$\sqrt{Hz}$ at 10 K are obtained without any external static field [7]. A force sensitivity in the order of 9x10^{-18} N/$\sqrt{Hz}$ should be reached at 4 K in a static magnetic field of 100 mT.

In this work we design, build and assemble the entire UHV machine working at a pressure of <10^{-10} mbar and at helium boiling temperature starting from the existing microscope and the Janis cryostat. This work took about one year producing hundreds of schemes and designs. The entire cryogenic machine plan is detailed in the appendix. For detailed subsystem schemes please refer to the scheme library in the appendices.

The extreme high sensitivity of 10^{-18} N/$\sqrt{Hz}$ that the magnetic force resonance microscope should reach, requires the study of interaction phenomena. The small spring constant for high force sensitivity makes it necessary to have the cantilever perpendicular to the sample surface. Otherwise, the cantilever will stick electrostatically to the sample surface. This vertical configuration introduces new design parameters involving the cantilever's approach to the sample. In fact the cantilever is subject not only on the lateral force gradient but also to a vertical force. The vertical attractive force as a uniform force will cause an increase in the frequency similar to the uniform gravitational force that causes a pendulum to have a frequency that is proportional to gravity.

The tip-sample interaction dissipation is then measured by the Q factor change as a function of the distance. The dissipation is caused mainly by the electrostatic charge fluctuation. The fluctuation of charge stored on a capacitance C induces the noise denoted as "KTC". The noise of the fluctuation charge is on the order of observed charge fluctuations of single-electron transistors. This shows a probably common origin of the charge fluctuation.

A severe loss in force sensitivity and a frequency shift are observed while exposing the cantilever with a magnetic tip to a homogenous magnetic field. The micrometer sized magnetic particles generate a magnetic field of 500 Gauss and magnetic field gradients (dB/dz>> 1x10^5 T/m). To minimize the damping losses of the cantilevers with ferromagnetic particles various magnetic materials (e.g. Sm_2Co_{17}, $SmCo_5$, $Nd_2Fe_{14}B$, and $Pr_2Fe_{14}B$) with different grain materials and domain sizes are investigated. The lowest magnetic dissipation is observed with $SmCo_5$ tips having a higher anisotropy constant. A correlation between frequency of oscillation and magnetic field hysteresis is then measured. A detection sensitivity in the order of $10^{-18} N/\sqrt{Hz}$ is reached at 100 mT. This sensitivity should be enough for measuring less than 100 electron spins.

Finally, a home-built spectrometer is compared with a home-built magnetic resonance force microscope with the sample mounted on the cantilever. At room temperature and at 50 mT the magnetic resonance force microscope has a sensitivity improvement of a factor of more than 100000. This suggests the huge potential of this instrument for biological and chemical sample analysis.

This work is part of ultimate limits of measurement of module IX of the National Center of Competence in Research in Nanoscience (NCCR). The NCCR is the national Swiss research projects in nano technologies with the leading house in Basel. The main goal of this submodule is to ultimately perform single spin experiments at low temperature and in ultra high vacuum (UHV). Achieving this goal requires mechanical force sensors to be improved and all relevant forces to be understood. The channels of energy dissipation should be determined in order to improve the detection sensitivity.

Outline of the thesis

The thesis is organized as detailed below. After a short introduction of scanning force microscopy, the beam deflection method and magnetic resonance force microscopy, the complete home built-system is presented in the second chapter. This chapter is focused mainly on the pieces designed by myself.

In the third chapter, a study of the autocorrelation and the Hilbert transformation is undertaken to determine the quality factor of a mechanical cantilever as an alternative method of the ring down measurement. This condition must to be met for maximal sensitivity.

In the fourth chapter a study of the magnetic tip and magnetic field interaction is explicated in order to understand the interaction and choose the best cantilever tip material. Theoretical models of the magnetic dissipation and frequency dependence are given.

The fifth chapter presents the measurement results of the magnetic tip as a function of the static magnetic field. The measurements compare the interaction of different tip materials, which have different anisotropy and size. The result shows a correlation between the magnetic anisotropy constant of the particle material tip and the static magnetic field.

The sixth chapter presents the tip-sample interaction of the long-range forces. The ultrasoft cantilever is not only sensitive to the lateral force gradient but also to the axial force. Moreover this chapter introduces the electrostatic force interaction and the fluctuation charge noise.

In the seventh chapter the magnetic resonance force signal and inductive magnetic resonance signal are presented theoretically.

The eighth chapter presents and compares the measurement performed with the continuous wave spectrometer and the room temperature magnetic resonance force microscope. The measurement shows an improvement of the signal-to-noise ratio of more than 100000.

Chapter 1

Magnetic resonance force microscopy

A short introduction in scanning force microscopy beam deflection methods and magnetic resonance force microscopy (MRFM) introduces the reader to the MRFM measurement technique. The MRFM technique inherits the sensitivity of the atomic force microscopy and has MRI capability.

1.1 Atomic force microscope and beam deflection

In a standard Scanning Force Microscope (SFM), the force between the surface and the microscopic tip is measured as a function of position. The longitudinal force gradient acting on the tip induces hardness or softness in the cantilever stiffness and a change of the cantilever frequency shift can be measured. The frequency oscillation of the mechanical lever is measured by a standard beam deflection method, but interferometer methods can be used with similar performance.

The layout of our actual magnetic resonance force microscope is represented in figure 1.1. An infrared superluminescent LED source is focused onto a cantilever and reflected into a split photograph detector diode. The ultrasoft cantilever is positioned perpendicular to the surface instead of the horizontal mounting seen in a standard Atomic Force Microscope (AFM). Otherwise, the low spring constant of the cantilever would bend it into contact with the sample due to the electrostatic force induced from charges between the surface and the scanning tip. In a perpendicular configuration the cantilever is not sensitive to the vertical force gradient, but only to the lateral force gradient induced by an asymmetry excitation in the sample. The lateral force gradient is measured using a standard FM-detection. This technique measures the frequency shift of the oscillating cantilever induced by an external force. Force and force gradient could be measured with the cantilever amplitudes changes. We use the frequency shift, because the dynamic response of a force gradient changes is faster than a static amplitude measurement.

An IBM ultrasoft cantilever tipped with an integrated micro-size permanent magnet material is operated in self-oscillation by a feedback loop (Control Oscillation Constant Amplitude: COCA) with nanometer amplitudes. In parallel, using a frequency counter, or a lock-in amplifier, the oscillation frequency is measured and the data are stored in a Labview program. The Labview program can shut down the feedback excitation with an electronic switch, whereupon the ring down signal can be acquired. Additional Labview modules have been introduced in order to determine the frequency shift and the spectrum of the signal with a precision of 50 mHz, for a long acquisition period. Moreover, with a homemade phase-lock-loop sub mHz precisions has been reached.

Since the piezo tube has a maximal extension in the micrometer range, the sample is approached manually with a stack motor from a maximal distance of 3 mm down to submicrometer distances, where the frequency of the cantilever begins to change due to the interaction with the sample. At the submicron range, the approach is undertaken more accurately with the piezo tube having a maximal extension of 1.5 micron per hundred volts. All the sample-tip distances are extrapolated from the maximum extension of the piezo tube. The ring down measurement is measured as a function of the distance of the sample from the tip and of the magnetic field.

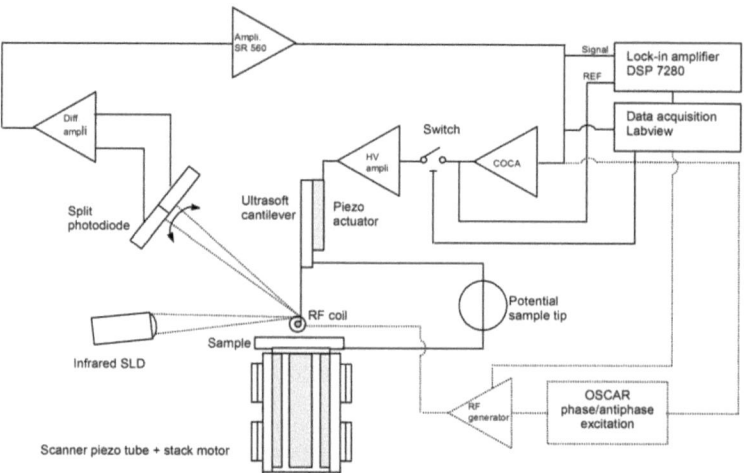

Figure 1.1: Magnetic resonance force microscopy setup. The picture shows the scheme used for measuring the tip-sample interaction, where the beam deflection method is used. The light from the super luminescent LED is focused on the cantilever, which reflects the light on a split photodiode. The differential detection is then amplified in the band width of the resonance frequency of the cantilever. The cantilever is then feedbacked-controlled by the Control amplitude Oscillator (COCA) that bears the Amplitude Constant. A lock-in amplifier is used to detect the real and imaginary part of the signal as the frequency. Simultaneously a frequency counter and a Labview program measure the frequency. The OSCAR module pulse method is indicated in dotted lines.

In this thesis two home-built magnetic resonance force microscopes were used. One operated at room temperature with a standard parallel configuration and in a vacuum of 5x10^{-6} mbar. This microscope is used for testing the cantilevers and performing the first continuous wave Magnetic Resonance Force Microscopy (MRFM) experiments. The other microscope operates at helium boiling temperature and in ultra high vacuum (UHV). This microscope is built for the ultimate spin detection, with a perpendicular configuration. The perpendicular configuration enables the use of an ultrasoft cantilever with spring constant of 0.1 mN/m and consequently improves the force sensitivity.

1.2 The Magnetic Resonance Force Microscope MRFM detection

The magnetic resonance force microscope (MRFM) uses the same principles as a standard magnetic force microscope (MFM) machine, which measures the magnetic interaction between the detection tip and the sample. The magnetic tip of a MFM machine usually comprises evaporated layer of iron (Fe) or Cobalt (Co) on the pyramidal Nanosensor cantilever tip. In MRFM usually a hard micro size magnetic material of $SmCo_5$ is used, because its strong anisotropy reduces the further oscillating field that causes a decoherence and a shortness of the relaxation time. The MRFM tipped cantilever induces gradients of more than 500000 T/m and magnetic field of 0.5 T at micrometer distances. Additionally, the MRFM has an alternating Radio Frequency (RF) field exciting the sample in order to excite the electron spin resonance.

Thus, a cantilever with an attached magnet is brought near the surface perpendicularly oriented, and a tuned Radio Frequency (RF) coil is positioned in close proximity to the end of the cantilever. In this way the cantilever is less excited by the radio frequency modulation. It is necessary to have the cantilever perpendicularly oriented, because of its low spring constant of 0.175 mN/m. Otherwise, the cantilever will stick to the sample surface via electrostatic forces. A scheme of the setup is presented in figure 1.2.

A triggered frequency RF pulse is applied, which changes the magnetization of the sample. Inversion of spins in this way is a resonant effect, and the large gradient from the tip confines this resonance to a thin slice of sample (absorption line width), depicted in yellow color. The interaction between the magnetic tip gradient and the sample spins induces a frequency shift measurable using by a standard magnetic force microscope (MFM). In order to a achieve detection sensitivity of 10^{-18} N/$\sqrt{Hz}$ the measurement should be performed under UHV conditions and at low temperature.

The full symmetry of the system reduces the lateral force to zero in a homogenously magnetized planar sample. The spins are excited in the ultimate sensitivity experiment of Rugar with the OSCAR protocol [4],[8]. The spins are consequently excited in phase or antiphase with the cantilever oscillation, which induces an asymmetry causing a lateral force and consequently a frequency shift.

In a single magnetic resonance force experiment many interaction forces and dissipation phenomena should be understood. The potential produced by the different

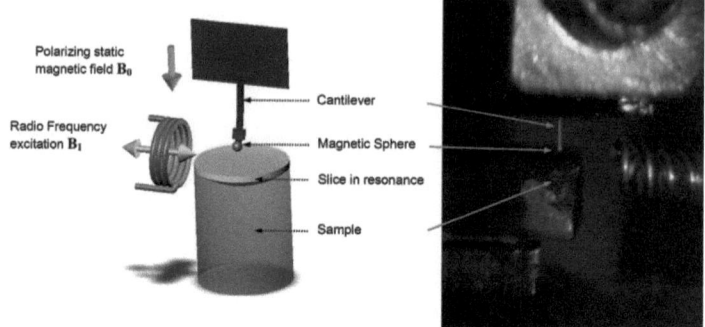

Figure 1.2: Magnetic resonance force microscopy sensor: Sketch and picture. The draw represents the setup used to measure the force signal from an excited sample. A polarizing field is applied parallel to the cantilever axis and a radio frequency excitation through the width of the cantilever. The strong magnetic tip gradient confines the resonance of the spins in a thin slice. By using the OSCAR protocol for spin excitation, a left-right imbalances of the spins was induced and a frequency shift measured. On the left side the photograph shows the cantilever, the coil and the sample.

work functions of the magnetic material and of gold-silicon induces a surface charge distribution. An electrostatic attractive force is induced by the surface charges, which can be reduced by compensating the contact potential. In a standard MFM with horizontal configuration the contact potential corresponds to the higher frequency, while in the vertical configuration, it corresponds with lower frequency. The vibrating cantilever produces a fluctuation field, which causes an energy loss measurable by a Q factor change in the cantilever.

Moreover, we report a severe loss of force sensitivity, when the cantilever is introduced in a homogenous magnetic field. A study of different magnetic materials shows a correlation between anisotropy and force sensitivity loss. The main advantages of this setup is a higher stability and sensitivity of forces gradients, due to the minuscule spring constant.

Chapter 2

Design and construction of a magnetic resonance force microscope

In order to achieve the ultimate limits of spin detection, we have built a low temperature ultra high vacuum (UHV) machine. The machine works in a wide of temperatures ranging from 4 K to 400 K. The existing microscope [7],[9] and the cryostat of Janis are integrated and tested. This chapter summarizes the choices and explains the different parts that constitute the entire machine.

2.1 Motivations

The goal of the magnetic resonance force microscope group in Basel is to build a machine with single electron spin sensitivity. The force induced by a single electron spin is on the order of aN (10^{-18} N). These forces are consequently more than 100000 times smaller than the usual forces measured by a standard atomic force microscope.

Ultrasoft non commercial cantilevers are used in order to reach this sensitivity. The cantilevers, due to the extreme soft spring constant, are extremely sensitive to external mechanical vibrations, which can prejudge the measurements. Therefore, the machine includes a 3 axis active damping table and an eddy current damping system, which guaranties mechanical isolation.

The thermodynamic vibration of the cantilever is reduced when the temperature is decreased. It is necessary to cool the cantilever to the 4.2 K temperature, because variations of temperature can be better controlled. In fact, at 4.2 K temperature, variation of less than 1 mK can be easily controlled, while at room temperature the variation of 1 K is difficult to control. Stability and reduction of the thermal noise consequently increases the sensitivity of the cantilever.

It is well know that a stable and ultra clean environment reduces the parameters, which can adversely the measurement. The implemented manipulators permit to prepare and transport the sample in the measurement region under UHV condition. Well defined and cleaned sample structures could in fact aid in a breakthrough in the MRFM technologies. The ultra high vacuum machine complicates the design resulting in the use of manipulators and an external lift.

The lift designed by Mewasa AG[1] allows to change the sample and cantilever under UHV condition and moves the microscope in the cryostat.[2] The cryostat is equipped with a split pair magnet which generate a magnetic field up to 7T and four windows in order to allows the optical access to the microscope. UHV conditions, high field and low temperature need particular materials with the following conditions: Low vapor pressure, non-magnetic characteristics and adequate thermal conductivity.

2.2 The Ultra High Vacuum (UHV) machine

The concept of the MRFM machine starts from the inherited microscope and cryostat. The system is designed around these two fundamental pieces and its current state is represented in figure 2.1. The system can be separated into three main regions: The measurement region, the preparation region and the load-lock region.

The measurement region includes the sample rod with the microscope, the middle chamber, the top chamber, the ion pump[3], the turbo pump[4] and finally the transfer line access through which it is possible to introduce samples and cantilevers. All these parts must be compatible with UHV conditions and a high magnetic field of 7 T.

The preparation region designed for sample and cantilever preparation works under UHV conditions. It contains a heater, an ion pump and a turbo pump. The aim of this chamber is to prepare both cantilevers and samples before they are introduced into the microscope. The heating anneals the crystal, removes contaminants from the surface and reduces the defects of the cantilevers. This process improves the quality factor by a factor of 10.

Finally, the aim of the load-lock is to introduce both the sample and the cantilever in the preparation chamber. Working in this region is delicate because it connects the UHV region with external pressure.

All these components are mounted on an active damping table designed by JRS[5]. The mechanical table isolates the microscope from the building and floor vibrations in a bandwidth between 1 to 60 Hz. During the measurement only the ion pump works in order to maintain UHV conditions (10^{-11} mbar), and to reduce the mechanical vibrations.

An overview of the UHV system is shown in figure 2.2. Besides the components

---

[1]Mewasa AG, Butz, CH - 8887 Mels, http://www.mewasa.ch

[2]Janis Research Company, INC, 2 Jewel Drive, P.O. Box 696, Wilmington, MA 01887-0696, USA, http://www.janis.com. Cryostat is a 7 Tesla vertical split superconducting magnet with a homogeneity of ± 0.1 % over a cm diameter sphere. The magnet includes a cancelation coil to reduce the field to less then 1000 Gauss. Data from Janis report.

[3]Varian, Vacuum Technologies, Via F'lli Varian 54, IT – 10040 Leini (TO), Vacion plus 500 (500 l/s), http://www.varianinc.com.

[4]Varian, Vacuum Technologies, Via F'lli Varian 54, IT – 10040 Leini (TO), Turbo-V 2000HT (2000 l/s), http://www.varianinc.com.

[5]JRS, HWL Scientific Instruments, Gmbh Georgstrasse 11, DE - 72119 Ammerbuch, http://www.hwlscientific.com

presented, the machine counts more homemade pieces such as the optical feed-trough, the radio frequency (RF) cable[6] and the shutter. The homemade optical feed-trough allows the introduction of infrared light into the machine. The homemade coaxial cable is designed to reduce the thermal conductivity and optimized to transfer RF power into the high temperature gradient zone (the performances of the cable are reported in the annexe).

Figure 2.1: Machine photograph. The photograph shows the middle chamber, the top chamber, the lift and the ion pump.

2.2.1 The measurement region

The measurement region, as introduced previously, includes the middle chamber, the top chamber and the sample rod. The top chamber and the sample rod are connected to the external lift and they can consequently be moved up and down. This complex construction has been required in order to transfer an annealed sample into the microscope and then inside the cryostat, under UHV conditions. The microscope knows two main positions: the transfer position, where a cantilever and a sample can be transferred and the measurement position, where the microscope is in the homogenous field of the superconductive magnet. In the measurement position the microscope is cooled by thermal conduction. The cone at the end of the sample rod touches a cold finger (the counter-cone) which is cooled with liquid nitrogen or helium.

---

[6]The transmission loss power of our homemade coax cable is 6-7% loss per m (value in the average of the best coax cable in commerce).

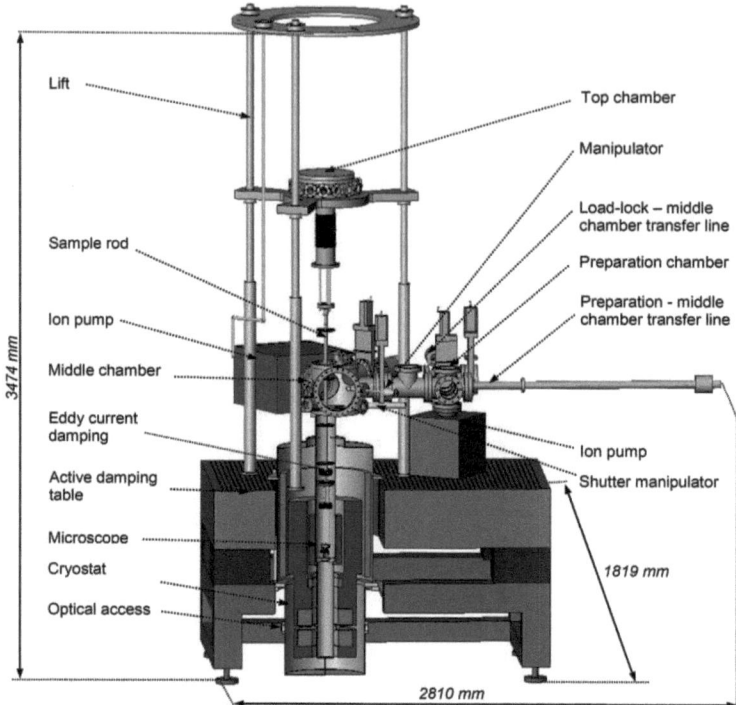

Figure 2.2: Overview of the magnetic resonance force microscope. The temperature can be varied between 3 K and 400 K. The bipolar split magnet can reach 7 T. The entire system is mounted on a 3D active damping table for reducing the external mechanical noise vibrations. An lift can move the microscope from the cold region to the middle chamber. In this position the cantilever and the sample change take place.

The cooling heat flow is guaranteed by many copper brads connecting the cone and the microscope. The microscope hanging on an eddy current damping system can be cooled down to a stable temperature of 5.5 K without pumping in the helium line.

Middle chamber

The middle chamber is connected with the preparation chamber and with the neck of the cryostat. In the middle chamber the change of the cantilever and of the sample takes place in this region. Many instruments are also connected to this piece: the turbo pump, the ion pump, the transfer line, some manipulators, the carrousel, the shutter, and the pressure gauge. This explain the reason why, the design has required many months of reflection in order to place the flanges in an optimal way without interfering with other machine parts inside and outside the chamber.

The main idea of the middle chamber is to be as compact as possible with the purpose of reducing the pumping time, to have the maximal access view and to have the best manipulators accessibility to the carrousel and to the microscope. The middle chamber is represented in figure 2.3.

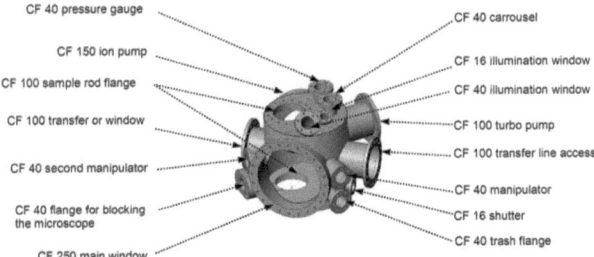

Figure 2.3: The middle chamber. The figure represents the middle chamber which was home designed and built by Mewasa AG. The material is a non-magnetic stainless steel 304. The flanges disposition are placed for compactness of the system and to simplify the internal manipulation.

The block scheme 2.4 shows by tasks the different functionalities that the middle chamber has afforded. The internal surrounding group represents the tasks, which have a strong impact on the middle chamber design. The flanges are placed to exclude any intersections of the instrumentation connected to the main chamber, to ensure a good mechanical stability and to ensure the functionality. The second surrounding group represents the tasks which does not involve any geometrical special conception.

The final design is a cylindrical geometry of non-magnetic stainless steel 304, with a height of 255 mm. The entire middle chamber flanges included has a maximal height of 330 mm and a maximal diameter of 440 mm.

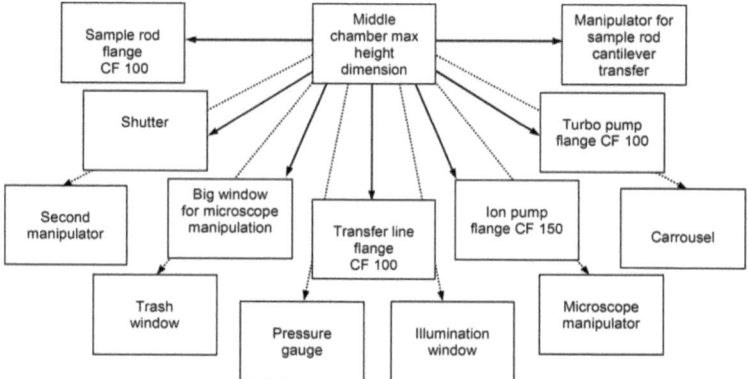

Figure 2.4: The block diagram shows all functionalities of the middle chamber. The first layer represents the defined important tasks that the middle chamber must afford, and the second the secondary tasks.

Top Chamber

The top chamber is not as complex as the middle chamber, nevertheless it plays a central role. All connectors such as the RF feed-through, the optical feed-through, and all electronic connectors[7] are included in the top chamber. The final design, represented in the figure 2.5, is a cylinder of stainless steel 304 of 110 mm of height and a diameter of 396 mm.

Figure 2.5: The top chamber. All cables are connected to this chamber and are introduced in to the machine by 16 feed-through. Stocked inside this chamber are between 2-3 m of optical fiber. This chamber is directly connected to the lift and to the sample rod.

The top chamber stocks 1-2 m of optical fiber. This stock can save precious time when the fibers has to be changed in the microscope region. A wide opening top flange is introduced for helping with the cabling and hand working inside the top chamber. Five connectors with 10 pins each, one RF feed-through and one

---

[7]Amphenol connector series, http://www.amphenol.com.

home made optical feed-through have been connected and they represent all the communication between the microscope and the external instrumentation.

Sample rod

The sample rod is the piece that connects the top chamber and the microscope. This piece guides all electrical cables, the optical fiber and the RF cable from the top region to the bottom region. An overview of the sample rod is show in figure 2.6.

Hence, four regions constitute the sample rod: the low temperature region including the microscope and the cone, the gradient temperature region including the cone and the sample rod, the filter vibration region including the eddy current damping system and finally the room temperature region including the end of the sample rod and its top.

The sample rod is submitted to a high gradient of temperature varying from 300 K to 4 K within a meter, while the cone is cooled by contact at helium boiling temperature. Therefore, the materials and the geometry have been optimized in order to reduce the losses. The microscope cooled down by 10 copper brads is hanging on the eddy current damping system, which ensures mechanical insulation for a frequency larger than 10 Hz.

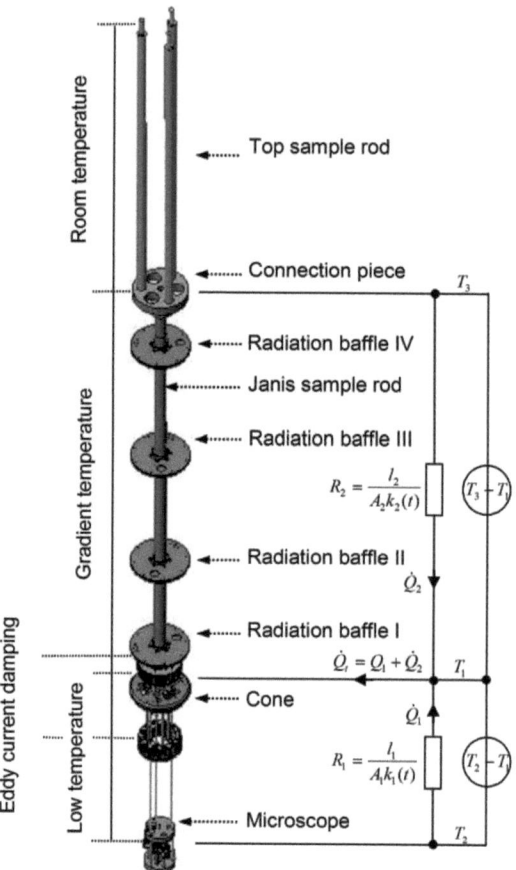

Figure 2.6: The sample rod. The figure shows the complete sample rod. This piece is subjected to a strong gradient of temperature. The sample rod isolates the microscope from temperature variation and from mechanical vibration. The systems have two thermal potentials causing the flux of heating. To reduce the heating flux causing the loss of power, the thermal resistor R_2 must be maximized. Conversely, the thermal resistor R_1 must be minimized in order to conduct the maximal cooling power to the microscope. The total cooling power is given by the contact surface between the cone and the counter-cone. So the cables, the sample rod and the RF cables set the minimal temperature reachable. The cooling power can be increased by pumping the helium. The eddy current damper mounted on the cone reduces the external vibrations, dissipating the energy by the eddy current. The transfer function of the eddy current damper begins from a few Herz and increases linearly with the frequency.

2.2.2 The cooling process

The cooling down process works by conduction and it is extremely influenced by the materials chosen. Over the atmospheric pressure the sample rod applies a force between the cone and the counter-cone ensuring by contact an optimal flow of the cooling power, as shows in figure 2.7.

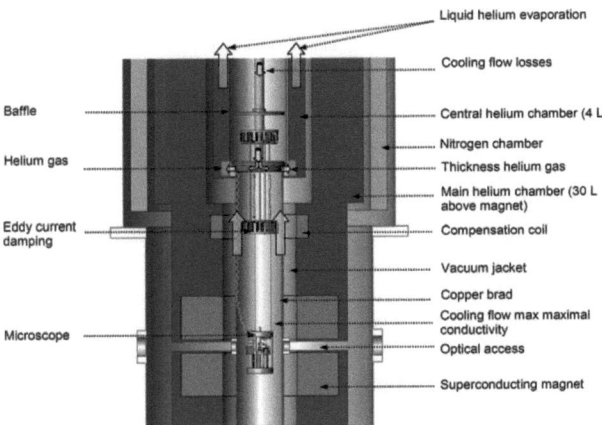

Figure 2.7: Side view of the cryostat. The picture shows the heat transfer from the microscope and from the sample rod to the cone. The heat is used to evaporate the liquid helium that is then evacuated. The cooling power can be increased by pumping in the helium line.

The heat flowing from the microscope and from the sample rod cause an evaporation of liquid helium in the central chamber of the cryostat, which causes the cooling down of these parts. At steady state conditions, the sample rod is consequently exposed to a strong temperature gradient. Over meter distance the temperature gradient varies from 300 K, the standard room temperature, to 4 K in the experimentation area. Under UHV conditions the main sources of heat losses are the heat conduction and heat radiation. In fact, the convection process can be neglected under UHV condition.

The conduction transport effect

At steady state condition the cooling power due to the heat conduction can be calculated. The cooling power efficiency is strongly dependent with the material employed. In order to reach 4-5 K the geometrical and the material characteristics are consequently optimized by calculating the heat conduction. The heat conduction is calculated using the following relation:

$$\dot{Q} = \frac{A}{l} k \left(T_2 - T_1 \right) \tag{2.1}$$

where A is the section surface of the corpus [m^2], l the length [m], K the thermal conductivity [Watts/m/K] and finally T_1 the temperature of the hot part [K] and T_2 the temperature of the cold part [K]. The relation 2.1 can be compared to the Ohm law, where the heat conduction corresponds to the current, the thermal conductivity corresponds to the electrical conductivity and finally the difference of temperature corresponds to the potential. A thermal scheme of the heat flow is represented in figure 2.6.

In order to optimize the heat conduction relation, the material and the geometry chosen for cooling down the microscope are related to the thermal conductivity and its specific heat. Usually, at helium boiling temperature a good conductor has the thermal conductivity 50 times higher. The thermal resistance R_1 (defined in figure 2.6) is consequently minimized choosing high thermal conductivity material and large section area. To reduce the cooling power loss, R_2 (defined in figure 2.6) is chosen as large as possible. Consequently low thermal conductivity material and small section area are chosen.

The figures 2.8, 2.9 and 2.10 represent the thermal conductivity, the electrical resistivity and the power transfer of different metal materials as a function of the temperature. The power transfer represents the integral of the thermal conductivity between 4 K and the temperature T for a conductor with one mm^2 section area and one meter of length.

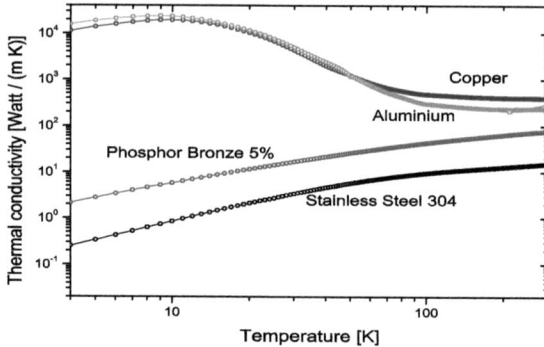

Figure 2.8: Thermal conductivity. The graph represents the thermal conductivity vs the temperature. At low temperature a good conductor has a good thermal conductivity.

In order to reach 4-5 K the thermal resistances are identified and optimized. We can define R_1 and R_2 as following:

R_1 includes 50 copper cables with 280 μm diameter, 10 copper brads with a section area of 5 mm² and a coax thermalized on the copper cone. The copper material is chosen because it has a very high thermal conductivity and low resistivity at 4 K.

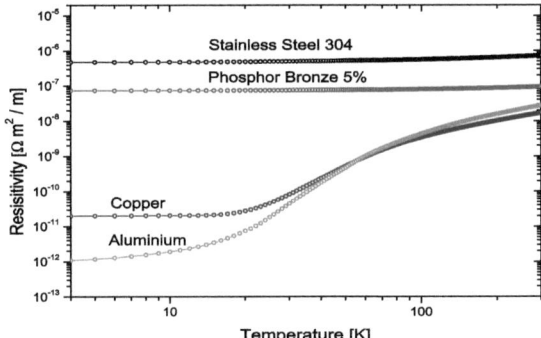

Figure 2.9: Resistivity. The graph represents the resistivity vs the temperature. The choice is complicated in choosing the cables that have a small resistivity and a small thermal conductivity (middle graph). The choice is moved to the phosphor bronze 5% material cable, which has a bad thermal conductivity and a good resistivity at helium boiling temperature.

R_2 includes 50 phosphor-bronze 5% cables with 280 μm diameter, a homemade coax and the sample rod. The electrical cables are in phosphor-bronze 5%, because this material meets a low resistivity and a low thermal conductivity. The RF coax cable is home made in order to reduce the thermal conductivity and optimize the transmission (please refer to annex A.2). The sample rod consists of a 0.5 mm thick stainless steel 304 pipe with an external diameter of 19.05 mm. This dimension thickness optimized the power losses without compromising the mechanical stability.

Table 2.1 shows that the total estimated power losses is of 0.5 Watts. The power loss is the sum of the power losses due to the phosphor-bronze 5% cables, the homemade coax cable and the sample rod. This power loss causes an evaporation of helium[8] equal to 0.7 l per hour. Currently, the largest power loss is caused by the

		Normal Boiling Point NBP [k]	Latent Heat of vaporisation at NPB [J/g]	Amount of liquid evaporated by 1 Watt at NPB [l/h]	Liquid density [g/cm³] at NBP	Gas density at NBP [g/cm³]	Latent heat density at NBP [J/cm³]	Cost for liter [CHF/l] at 2003
8	helium	4.22	20.9	1.38	0.125	1.66×10^{-4}	2.61	~ 14.5
	nitrogen	77.3	198	0.024	0.808	1.16×10^{-3}	159.9	~ 1.0

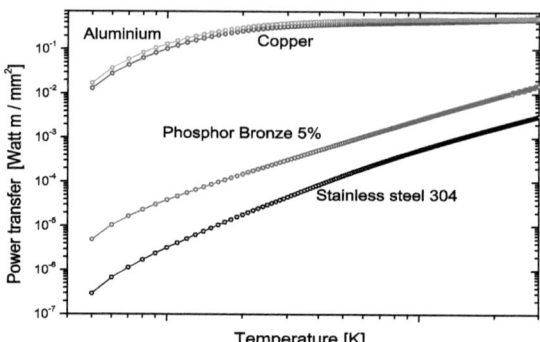

Figure 2.10: Power transfer. The graph represents the resistivity vs the temperature. The graph represents the real power transfer at steady state condition caused by a cable one meter long and with a section area of one mm².

	Material	Number	Length [m]	Dimensions	Power losses [Watts]
Cables	Phosphor-Bronze	50	1	⌀ = 280 μm	0.059
	Copper	50	1	⌀ = 280 μm	1.863
Coax Cable	Stainless steel	1	1	⌀ =3 mm; ⌀=11 mm x 1.5 mm	0.324
	Copper	1	1	⌀ = 0.51 mm; ⌀=2.2 mm x 0.26 mm	1.082
Sample rod Janis	Stainless Steel	1	1	⌀ = 19.05 mm x 0.5 mm	0.111

Table 2.1: Power losses due to thermal conduction. The table shows the real power losses caused by the different elements. The home build coax cable is compared with a commercial coax

coax cable. However, the home built coax cable has a low energy loss if compared with a commercial cable[9] as represented in table 2.1. The section dimension should be reduced from the actual 52 mm^2 to 17 mm^2, by reducing the wall thickness[10]. This improvement should reduce the power loss by a factor of three, that is to 0.105 Watts.

The temperature reached with the power loss of 0.275 Watts (new coax cable design) should be of 4.6 K. The temperature is calculated with the formula 2.2, which relates the cooling power of the machine and the final steady state temperature.

$$T_2 = \frac{P^2 \frac{\rho_{gas}}{\rho_{liquid}}}{L_{heat\ density} S^2 k} + 4.22 \quad (2.2)$$

where P is the power loss, ρ_{gas} the density of the gas, ρ_{liquid} the density of the liquid, $L_{heatdensity}$ the latent heat density, S the surface and K the thermal conductivity.

The radiation transport effect

Beside the heat conduction the heat radiation is the another mechanism which transports thermal energy. This process does not need a medium to propagate, because the energy is transported by electromagnetic waves. Thermal radiation can be expressed by:

$$\dot{Q} = e\sigma A(T_1 - T_2)^4 \quad (2.3)$$

where e is the emissivity (e=0 for a mirror and e=1 for a black surface), σ the Stefan-Boltzmann constant (5.68×10^{-8} Watts/m^2/K^4), A the section surface and T_1 and T_2 the temperature of the hot and cold medium.

From the equation 2.3 appears that this effect does not depend on the distance, but only on the section area and on the difference of temperature. Figure 2.11 shows the power dissipation between two reflexives surfaces (e=0.1) of 1 cm^2, one at 4 K and the other at T temperature.

In order to reduce the losses the baffles surface[11] are polished and gold plaited. The baffles are also anchored and thermalized to the internal pipe of the cryostat. The baffles are consequently cooled down to a low temperature by the VTI. Finally, the first radiation baffle is placed as close as possible to the cone in order to the reduce the temperature difference. The figure 2.12 shows the improvement.

The temperature difference reached is 20-30 K, as shows in figure 2.12. This temperature difference causes a power loss of 10^{-5} Watts/cm^2. This value is extremely small if compared to a power transfer of 0.01 Watts/cm^2 caused by a temperature difference of 100 K.

---

[9]Coax EZ-86-CU-TP-M17, Huber+Suhner AG, Verkauf Schweiz, Tumbelenstrasse 20, Ch-8330 Pfäffikon ZH http://www.Hubersuhner.ch.
[10]This action could compromise the RF characteristics
[11]The surface covering is of 90% of the internal VTI tube.

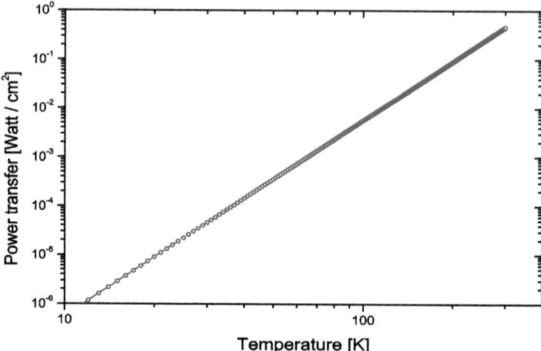

Figure 2.11: Radiation power transfer as a function of the temperature. The graph represents the heat power transfer as a function of the difference of temperature between two surfaces. It should be remarked that the radiation power transport does not depend on the distance. This effect can be neglected, if the difference of temperature between the two surfaces is smaller than 70 K. In this range the power transfer is equal to 1 $[mWatts/cm^2]$.

2.2.3 Sample rod and mechanical noise filtering

Magnetic resonance force microscopy is a mechanical detection, and for this reason all mechanical noise vibrations should be reduced and filtered. The eddy current damper is chosen in order to attend this goal.

The eddy current damper is a passive methods and during a mechanical vibration the copper pieces vibrate and dissipate the energy. The copper pieces vibrate in fact in a magnetic field generated by 24 magnets of Sm_2Co_{17}. Consequently, the vibrations induce the eddy current which dissipate through the heat of the copper material the energy. The vibration is then reduced (see figure 2.13).

The eddy current damper mounts 24 magnets, which can strongly interact with the magnetic field generated by the cryostat. Consequently, the eddy current damper is placed in the compensation magnetic field region, where the maximal magnetic field is of 100 mT. This magnetic field it is not enough to depolarize the magnetization of the 24 Sm_2Co_{17} magnets, which have a demagnetization field of 0.62 T. The gradient generated by the superconductor magnet in the compensation coil region interact with the magnets and can produce strong forces up to 1 N and torques up to 200 N/m. This forces can cause serious damages to the internal system. For this reason, the magnets are encapsulated in an aluminum frame, which not only guarantee mechanical fixation but also an easy replacement.

Moreover, in order to decrease the interactions when the magnetic field is swept the magnets are mounted with the polarization parallel and anti parallel to the axis of the superconduction magnet. An overview of the damper is shown in figure 2.13.

The eddy current damping system is a mechanical filter and consequently its has

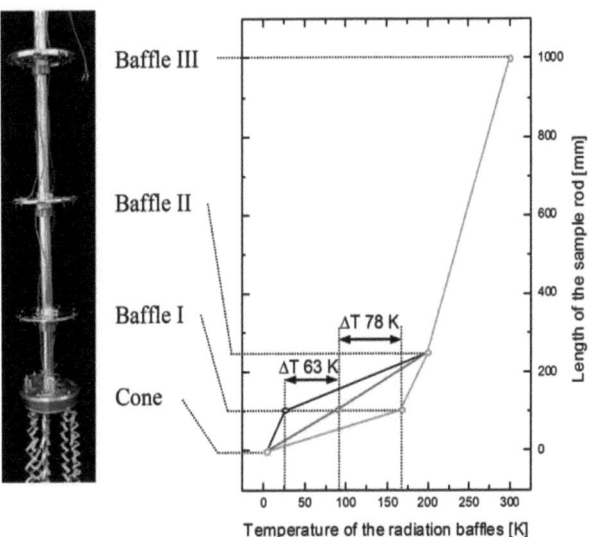

Figure 2.12: Microscope and baffles temperature. The picture on the left shows the realized sample rod. On the right the graph shows the temperature reached on the microscope and on each baffles. The black line shows the temperature distribution of the sample rod without CuBe springs and with a cross section of the radiation baffle 66% of the sample tube. The red line shows the temperature distribution of the sample rod with CuBe springs and with a cross section of the radiation baffle 66% of the sample tube. The green line shows the temperature distribution of the sample rod with CuBe springs a with a cross section of the radiation baffle 95% of the sample tub

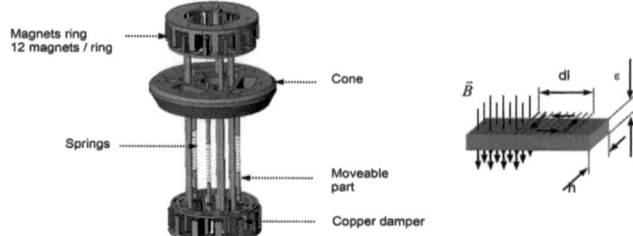

Figure 2.13: The eddy current damping system. The picture shows an overview of home built eddy current damping system. 24 magnets have been fixed in an aluminium frame reducing the risk of internal damage. Over the magnetic gradient and the change of temperature a magnet can detach from the support or brake. On the right a simple explanation of the eddy current principle is provided. The force depends on the square of the magnetic field and linearly with the speed.

a resonance frequency, which should be in the bandwith of the damping table (from 1 to 50 Hz). This guaranties that external noises do not excite the damper.

The eigenfrequency of the eddy current damping system is calculated to be 3.5 Hz. This frequency is calculated by knowing that: Three springs of 13 mm length with a spring constant of 92 N/m support a microscope mass of 362 g and an moveable part of the eddy current damper of 208 g.

The mechanical eigenfrequency is then measured exciting the system with white noise. The signal is measured with Labview and the Fast Fourier Transformation (FFT) plotted. The first vertical eigenfrequency of the mechanical system is 3.5 Hz at room temperature and 6.8 Hz at nitrogen boiling temperature. An unexpected resonance frequency is measured at 1 Hz at room temperature and at 0.2 Hz at 77 K, as shown in the middle picture 2.10. Finally the third graph of picture 2.10 represents the transfer function. The transfer function is measured by exciting the eddy current damper with a mechanical transducer at different frequencies and at fixed amplitude. The transfer function shows a low pass filter at 1 Hz.

In order to understand the relation between the number of magnets, which increase the total amplitude field and the damping friction, we have calculated the force. The damped force is calculated with the current generated by the electric potential induced by the variation of the magnetic flux in the copper pieces. For simplification we suppose an average magnetic field of 0.015 T between the magnets. The force extrapolated is given by the formula 2.4.

$$F_t = |idl \times B| = 24\frac{B^2 h^2 t}{\rho}\dot{x} \qquad (2.4)$$

where B is the average magnetic field, h the width of the copper piece (10 mm), ρ the resistivity of copper (1.72×10^{-5} Ωm), t the thickness of the copper piece (3 mm) and finally F_t the total force generate by the 24 magnets. The eddy current damper is tested using a home built two quadrant detection system and the signal of the vibration measured by a Labview program. The eddy current damper is excited at the eigenfrequency and the decay time measured after the excitation signal is abruptly stopped. Therefore, the magnets are introduced in the frame and the experiment repeated. The data are represented in figure 2.14, 2.15, 2.16 and fitted with the approximate equation 2.5. The variation of the decay time is correlated with the losses.

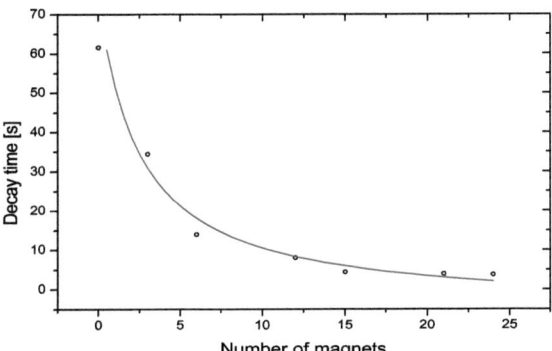

Figure 2.14: The eddy current damper decay. The graph represents the decay time of the eddy current damping vs the magnet introduced. The curve is fitted using equation 2.5.

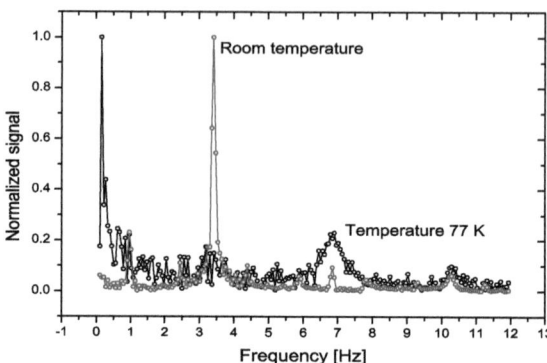

Figure 2.15: The spectrum of the excited eddy current damping. The graph shows the spectrum of the excited eddy current damping system. A resonance frequency of 3.5 Hz is measured at room temperature and it increases to 6.8 Hz at 77 K.

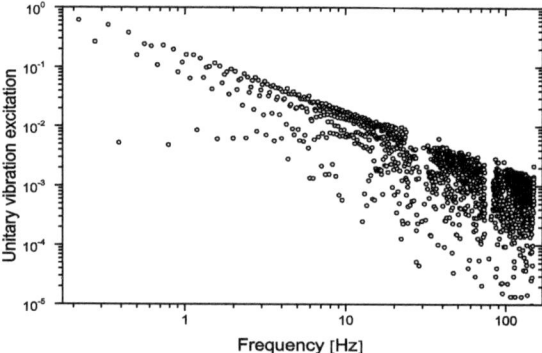

Figure 2.16: Transfer function of the eddy current damping. The graph shows the transfer function of the eddy current damping at room temperature.

Chapter 3

Cantilevers: Thermodynamic equilibrium and quality factor

In order to detect signals in atto Newton (10^{-18} N) range, the mechanical detector has to be in thermodynamic equilibrium. Due to the extreme small cantilevers stiffness (0.1 mN/m), the cantilevers are strongly perturbed by the mechanical noise.

In order to reach the thermodynamic equilibrium, the amplitude noise vibrations have to be measured and reduced. The gaussian distribution and the autocorrelation provide the tools to measure the amplitude noise vibrations, which have to be reduced to the thermodynamic level.

At thermodynamic equilibrium the autocorrelation measure provides not only the thermodynamic vibration amplitudes but also the quality factor of the mechanical oscillator. The decay time of the autocorrelation function is in fact correlated with the dissipation processes of the lever.

3.1 Theory: Harmonic mechanical oscillator in thermodynamic equilibrium

The thermodynamic equilibrium is the limit of the cantilever sensitivity. To reach this equilibrium many efforts were made to reduce the external mechanical noise and electronic noise. The mechanical noise is reduced with an active damping table[1] and an eddy current damping[2] implemented in the MRFM machine. The electronic noise is reduced to a sub thermodynamical equilibrium level with two analog band filters and with the separation of the cables.

During the noise measurement any mechanical noise sources and electronic noise devices should be turned off. The turbo pump should not be operative, because it excites the cantilever and increases in this way the vibration noise. All measurements should be performed using the ion pump, which does not produce any mechanical vibrations.

---

[1]The transfer function of the damping table has a band filter from 1 to 60 Hz.

[2]The transfer function of the eddy current damping has a low pass filter at 1-2 Hz at room temperature.

3.1.1 The Gaussian distribution

The probability density distribution provides information about the coherent excitations and amplitude oscillations. In thermodynamic equilibrium the probability density distribution must have a Gaussian distribution. The Gaussian distribution proves that the mechanical cantilever is not subjected to although any coherent mechanical noises. Anyway any incoherent noises excitations could be presents.

The figure 3.1 represents the amplitude distribution of the measured 2.110 kHz IBM tipped cantilever in a thermodynamic state. The measurement is performed at room temperature. In order to produce this graph we record in a complete passive mode the 4-quadrants signal for a period of 60 seconds with a sample rate two times faster than the first eigenfrequency of the cantilever. Then the signal is divided in more equidistant intervals of amplitudes and the graph is finally normalized in order to have a probability area of 1.

When the cantilever is excited with a coherent mechanical signal, the density distribution enlarges and two peaks appears. The graph 3.2 shows the normalized amplitude density distribution as a function of different piezo excitations varying from 1 mV to 1 V. The turbo pump also excites the cantilever with a coherent mechanical noise excitation. This effect is measured by a broadening of the Gaussian distribution.

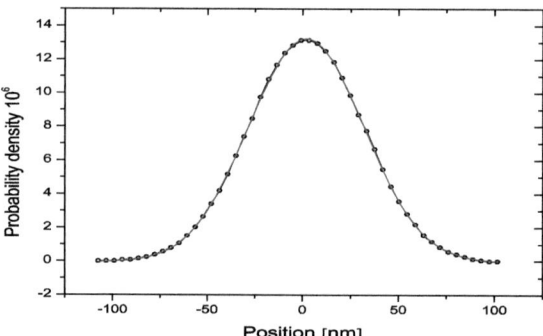

Figure 3.1: Gaussian probability distribution. The graph shows the amplitude probability distribution of the IBM cantilever in a thermodynamically state. The red curve is a Gaussian fit of the amplitude probability distribution, which has a width of 60.46 nm ±0.128 nm.

The average amplitude oscillation of the excited states is extrapolated by adjusting the probability distribution of the thermodynamic state. At room temperature we can use the equipartition theorem to calculate the square of the amplitude:

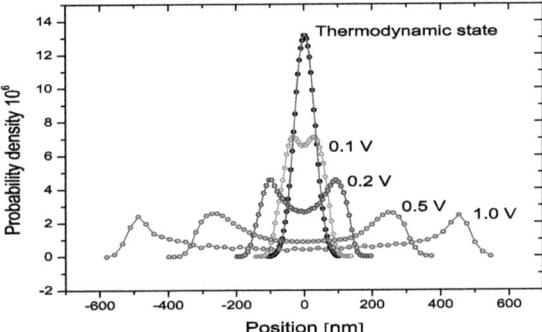

Figure 3.2: Probability distribution. The graph shows that different amplitude excitations are applied to the piezo excitation; more the excitation amplitude increases more the probability distribution split in to states.

$$<x^2> = \frac{k_b T}{k} \qquad (3.1)$$

where $<x^2>$ is the average of the square oscillation, k_b the Boltzmann constant and T the temperature. The square of the amplitude oscillation is then compared with the following relation:

$$<x^2> = \int_{-\infty}^{\infty} (cx)^2 \rho(x) dx \qquad (3.2)$$

where $\rho(x)$ is the density distribution and c the calibrating constant.

When the thermodynamic state is calibrated, it is possible to extrapolate the average amplitude of the cantilevers that are coherently excited. The graph 3.2 represents the amplitude oscillation as a function of the piezo excitation.

At room temperature, the thermodynamic amplitude of the IBM cantilever is 30 nm. When the cantilever is excited with a coherent excitation, the average amplitude increases linearly.

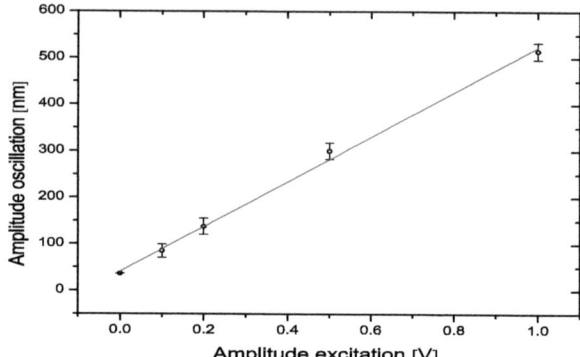

Figure 3.3: Amplitude oscillations. The graph shows the amplitude oscillations as a function of the piezo excitation. At thermodynamic condition and at room temperature the cantilever oscillation amplitude is calculated to be 30 nm. The cantilever amplitude oscillation increases linearly with the amplitude excitation.

3.1.2 The autocorrelation

The autocorrelation provides a powerful tool to measure the noise amplitude of correlated and uncorrelated noise. The autocorrelation of a signal U(t) as a function of the time delay τ is reported in [7],[10].

$$G(\tau) = <U(t)U(t+\tau)> = \lim_{T \to \infty} \frac{1}{T} \int_0^T U(t)U(t+\tau) dt \qquad (3.3)$$

At thermodynamic equilibrium the phase of the oscillation signal between U(t) and U(t+τ) are uncorrelated and the angle varies randomly between 0 and 2π. Consequently, by averaging of the product U(t)U(t+τ) the autocorrelation G(τ) must vanish with the time. If the autocorrelation function G(τ) does not vanish, the two functions U(t) and U(t+τ) maintain some correlation and their product maintain a constant value.

Figure 3.4 reports the mechanical noise vibration measured for 50 seconds with an IBM cantilever and figure 3.5 its autocorrelation. In a thermodynamic state the autocorrelation decays at infinity to 0. The right side graph shows this trend.

The quality factor detection

At thermodynamic equilibrium the autocorrelation provides the measure of the quality factor of the cantilever. The quality factor is the parameter that define the amplification at resonance frequency. In the literature, the quality factor of a mechanical oscillator is determinated mainly by three methods [11], namely: The

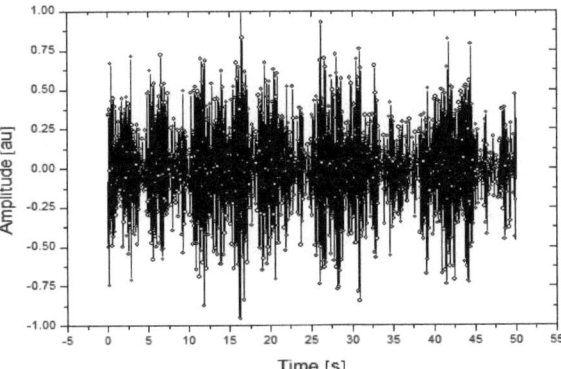

Figure 3.4: Thermal noise. The graph represents the four-quadrant detector signal measured without any excitations.

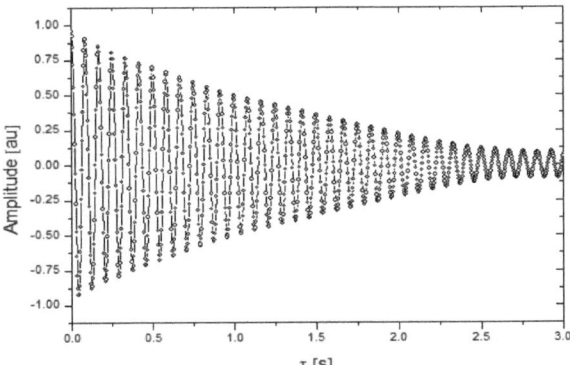

Figure 3.5: The autocorrelation. The graph shows the autocorrelation of an IBM cantilever in the thermodynamic state. The signal has like a memory and the phase can be predicted in the coherence time, after which period the signal has a random phase and the signal decays to 0.

ring down measurement, the power spectral density methods and finally by the methods based on the Brownian thermal motion of the mechanical cantilever.

The ring down method assumes that the cantilever oscillates at its natural resonance frequency at a constant amplitude. The feedback is then abruptly grounded and the decay recorded.

In the power spectral density methods, the power spectral signal is measured and from the spectral width of the resonance peak the quality factor of the harmonic oscillator deduced. This method has the disadvantage that at short distance between tip and sample the shape is not Lorentzian.

The Brownian motion methods is the most accurate methods for determine the Q factor on condition that it is very large. This methods extrapolates the Q factor by fitting the envelope of the autocorrelation. The envelope is calculated with the amplitude of the two orthogonal functions: the real and the Hilbert transformation.

The total amplitude of the real and imaginary signal is the envelope of the autocorrelation, which can be fitted with the exponential decay function. If the autocorrelation is the real function $G(\tau)$, the imaginary part is its Hilbert transformation. In the time domain the Hilbert transformation is defined by the convolution between the Hilbert transformer $1/(\tau\pi)$ and a function $G(\tau)$. The Hilbert transform of a function $G(\tau)$ is defined of all t by

$$\hat{G}(\tau) = \frac{1}{\pi} \int_0^\infty \frac{G(\tau)}{\tau - v} dv \tag{3.4}$$

The Hilbert transformation can be applied to the autocorrelation function measured previously. The results are two orthogonal functions, which are represented in figure 3.6.

The envelope of the autocorrelation is the amplitude of the complex number calculated from the real and imaginary part as shows in figure 3.6. Consequently, the envelope decay is fitted with the exponential decay function from which the Q factor is extrapolated. The main advantage of this method is the possibility to measure the quality factor without any excitation. So with the maximal sensitivity.

The Brownian thermal vibration method has the advantages that the cantilever vibrates at the smaller amplitude. In fact large amplitudes could strongly affect the quality factor measurement, because the cantilever is more exposed to inelastic processes.

We report that the Q factor measured with the IBM cantilever as shown in figure 6.1, agrees with the ring down measurement to within 12% of error for 5 measurements. The ring down measurement shows a Q factor of 30054 ± 1586, while the Brownian motion method shows a Q factor of 26821 ± 1232.

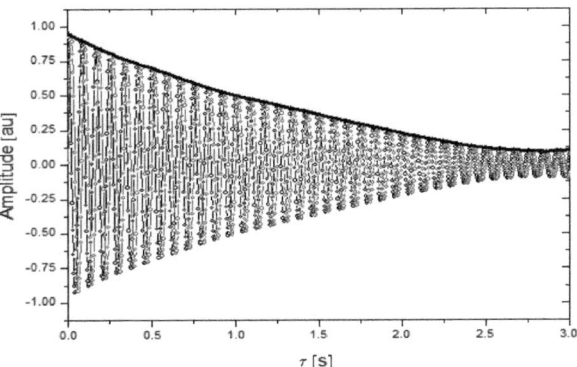

Figure 3.6: The autocorrelation and the Hilbert Transformation. The graph on black color is the autocorrelation function of the signal measured above and the graph on red color is its Hilbert transformation. These two functions are orthogonal and the total amplitude represents the envelope of the decay. The envelope can be fitted easily with an exponential decay function. A Q factor of 26821 is calculated. This value corresponds to the value measured with the ring down measurement with a 12% precision.

Chapter 4

Damping losses of cantilevers

In magnetic resonance force microscopy single spin experiment forces in the atto Newton range have to be measured. Non commercial, soft single crystalline silicon bar cantilevers with a high quality factor and minimized spring constants have to be used, in order to improve the detection sensitivity. In our low temperature force microscope we obtain force sensitivities on the order of 10^{-18} N/$\sqrt{Hz}$ at 10 K [7].

Micrometer sized magnetic particles, which generate a magnetic field of 500 Gauss and magnetic field gradients (dB/dz>> 1 G/nm) are attached on ultrasensitive cantilevers. A severe loss in force sensitivity and a frequency shift are observed while exposing the cantilever with the magnetic tip to a homogenous magnetic field. To minimize the damping losses of the cantilevers with ferromagnetic particles, various magnetic materials (e.g. Sm_2Co_{17}, $SmCo_5$, $Nd_2Fe_{14}B$, and $Pr_2Fe_{14}B$) with different grain and domain sizes are investigated. The lowest magnetic dissipation is observed with $SmCo_5$ and $Pr_2Fe_{14}B$ tips. We try to explain the dissipation effect of cantilevers with magnetic tips.

4.1 Determination of the tip mass

The determination of the mass glued at to the end of the mechanical resonator is well known in the literature [12]. Here, we did not calculated the change of the frequency, but instead the different eigenfrequencies as a function of the attached mass. The lower eigenfrequencies are more affected by the attached mass than the higher eigenmodes, and consequently the ratio between the eigenfrequencies changes. The first eigenmode is the most sensitive, because it varies strongly by increasing the mass glued at the end of the cantilever.

Starting from the motion equation, it is possible to add an inertia mass and then numerically solve the system of equations after defining the boundary conditions. The motion equation for a free vibration cantilever with homogenous cross section is well known in the literature and this differential equation of fourth order is [13],[14]:

$$EI\frac{\partial^4 \nu}{\partial x^4} + \rho A \frac{\partial^2 \nu}{\partial t^2} = 0 \qquad (4.1)$$

where E is the modulus of elasticity, ρ is the mass density, A is the cross section area and I is the area moment of inertia. One solution can be obtained by the separation of variables. The equation is then reduced to:

$$\Phi^{iv}(x) - \frac{\omega^2 \rho A}{EI}\Phi(x) = 0$$
$$\ddot{Y}(t) + \omega^2 Y(t) = 0 \tag{4.2}$$

The solution of the first equation is given by a sum of trigonometric and hyperbolic equations (Annex A.12), and can be used to numerically calculate the different eigenmodes. As show in the figure 4.1 the cantilever supports a mass at its end so the fourth boundary conditions lead to:

$$\begin{aligned}\Phi(0) &= 0 \\ \dot{\Phi}(0) &= 0 \\ EI\Phi''(l) - \omega^2 J\Phi'(l) &= 0 \\ EI\Phi'''(l) - \omega^2 m\Phi(l) &= 0\end{aligned} \tag{4.3}$$

The first two boundary conditions lead to the fixed point at x=0, the third from the moment applied at the end of the cantilever and the fourth from the transverse force. Figure 4.1 sketches the boundary conditions and the oscillating cantilever.

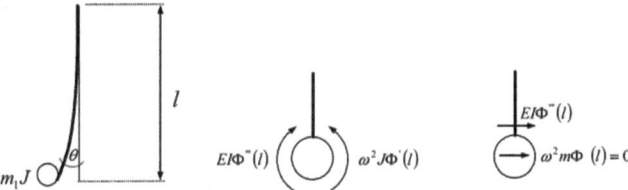

Figure 4.1: Tipped cantilevers. The figure shows the cantilever with a mass m_1 and an inertia J glued at its end. The boundary conditions have been applied at the extremity of the shape function. The parameters are: E the elastic modulus constant of silicon, I the section inertia, θ the angle of oscillation and ω the frequency of oscillation.

The numerical results of the IBM and Nansensors cantilevers are plotted as a function of the mass attached at the end of the mechanical lever shown in figure 4.2 and 4.3. The first eigenfrequency and the frequency shift can be also calculated using the point-mass model solution given by Rabe [12]. The ratio of the eigenfrequency of the Nanosensors cantilevers (represented in figure 4.2) and of the IBM cantilever (represented in figure 4.3) are numerically calculated, plotted and compared with experimental data. The tip mass is extrapolated from the frequency shift and the ratio between the two eigenfrequencies. In fact a variation of the ratio between eigenfrequencies is detected, when the mass is increased. The mass sensitivity for small masses of higher eigenmodes is comparable to the mass sensitivity of the fundamental mode.

The figure 4.2 and 4.3 show the trend of the frequency shift as a function of the mass attached to the end of the cantilever. The Nanosensor cantilever has a strong frequency variation starting at $10^{-10}kg$, as compared to $10^{-14}kg$ for the IBM cantilever. This strong difference is attributable mainly to the stiffness of the

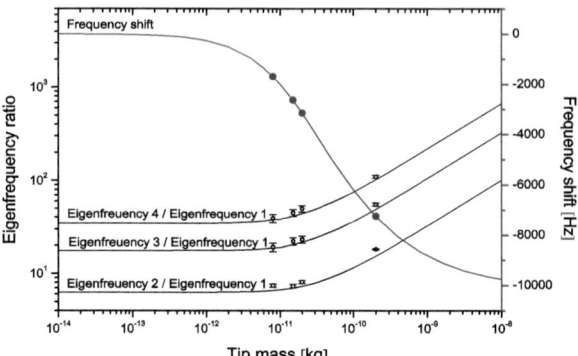

Figure 4.2: Nanosensor eigenfrequency vs mass. The graph represents the frequency shift of the Nanosensors cantilever with the first eigenfrequency calculated at 10160 Hz. The ratio between first, second, third and fourth eigenmode has a small change for masses smaller than 10^{-12} kg. The strong change is in the range of $10^{-12} - 10^{-10}$ kg. The ratio between eigenfrequency can be used for determine the mass glued, for masses larger than 10^{-10} kg.

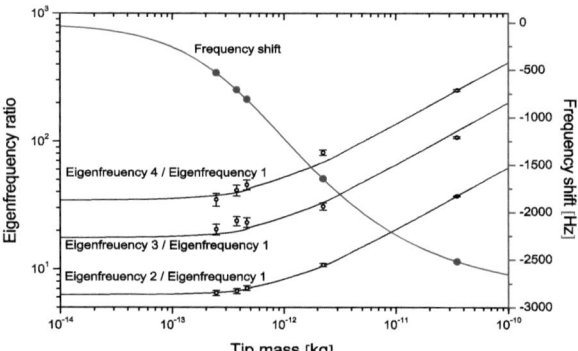

Figure 4.3: IBM eigenfrequency vs mass. The graph represents the frequency shift of the IBM cantilever with a first eigenfrequency calculated at 2703 Hz. The ratio between first, second, third and fourth eigenmode has a small change for masses smaller than 10^{-14} kg. The strong change is in the range of $10^{-13} - 10^{-11}$ kg. The ratio between eigenfrequency can be used for determine the mass glued, for masses larger than 10^{-13} kg.

mechanical levers. The ratio between the eigenfrequencies increases as the mass attached at the end of the mechanical lever is increased.

The strong change of the first eigenmode in both Nanosensor and IBM cantilevers suggests that the first eigenmode is the sensitive vibration mode. In fact, the cantilever has an internal elastics frictional processes, which are correlated with the frequency vibrations. The dispersion of energy results in a line broadening of the eigenfrequency defining a change of phase and a decoherent time in the elastic processes.

4.2 Damping losses of cantilevers in an external magnetic field

This chapter explains the frequency shift, the force sensitivity loss with hard and soft magnetic tip materials measured while exposing a tipped cantilever to a magnetic field. Finally, the magnetic frictional constant is extrapolated from the magnetic loss measured.

4.2.1 Theory: Interaction between magnetic tip and external magnetic field

The interaction between the magnetic particle and the homogenous magnetic field can be measured by the frequency shift induced on the mechanical lever. The frequency shift induced by the interaction with the static magnetic field can be calculated from the torque acting on the particle. The figure 4.4 sketches the interaction between tipped cantilever and external magnetic field. In order to calculate the torque acting on the cantilever the total potential energy of the mechanical resonator in an external magnetic field is minimized. The total potential energy is:

$$E(\theta, \theta_m) = E_{Zeeman}(\theta, \theta_m) + E_{Spring}(\theta) + E_{Anisotropy}(\theta_m) \quad (4.4)$$

The total potential energy of the mechanical oscillator can be calculated from three main terms [15]: The Zeeman energy term, the potential energy of the spring and the anisotropy term (we neglect the exchange energy). The Zeeman energy is :

$$E_{Zeeman}(\theta, \theta_m) = -M_s V B \cos(\theta - \theta_m) \quad (4.5)$$

where $\mathbf{M}_s$ is the magnetization, V the volume, $\mathbf{B}$ the magnetic field, θ the cantilever angle and θ_m the angle between the magnetization tip particle and the external magnetic field.

The energy potential of the cantilever is given by

$$E_{Spring}(\theta) = \frac{1}{2} k_0 (l\theta)^2 \quad (4.6)$$

where k_0 is the spring constant, l the length of the cantilever and θ the deflection angle represented in the figure 4.4.

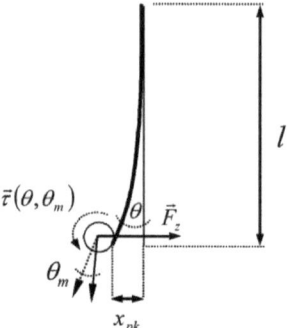

Figure 4.4: Tipped cantilever in a magnetic field. The sketch represents the cantilever with the magnetic particle subjected to the external magnetic field. The magnetic particle exercises a lateral force on the mechanical resonator, causing an increase in the resonance frequency. A momentum at the end of the cantilever will cause a reduction of the resonance frequency. The parameters are: l the length of the cantilever, x_{pk} the peak amplitude, F_z the lateral force, the θ angle of oscillation, m the angle of the magnetization rotation.

The energy shape anisotropy in a magnetic material where the grain size is smaller than the critical size of a single magnetic domain is [16]:

$$E_{Anisotropy}(\theta_m) = \mu_0 \left(M_s V\right)^2 \frac{D_{//} \cos^2(\theta_m) + D_\perp \sin^2(\theta_m)}{2V} \qquad (4.7)$$

where μ_0 is the magnetic permeability in vacuum, and $D_{//}$ and $D_\perp$ the principle demagnetization factor of a ellipsoid parallel and perpendicular with respect to the static magnetic field.

When the grain size has the same dimension as that of a magnetic domain the anisotropy term reduces to the magnetocrystalline anisotropy (values reported in table 4.1). In other words, the expression of the anisotropy depends on the magnetocrystalline symmetry. The simple anisotropy expression (4.8) is widely used, but sometimes it is necessary to take high order anisotropy constants into consideration [17].

$$E_{Anisotropy}(\theta_m) = K_1 V sin^2(\theta_m) \qquad (4.8)$$

The anisotropy term can be distinguished between the macroscopic shape anisotropy important in small aspherical particles and the magnetocrystalline anisotropy, which is an intrinsic lattice property. The single domain radius R_s is of order of 0.2 μm in modern permanent magnet. Consequently a magnetic particle larger than 1 μm is multidomain.

To calculate the anisotropy rotation angle θ_m we minimize the potential energy. The solution is calculated for the geometry anisotropy and for the magnetocrystalline anisotropy.

$$\frac{\partial E(\theta, \theta_m)}{\partial \theta_m} = 0 \qquad (4.9)$$

Cantilever	Nanosensor 1	Nanosensor 2	Nanosensor 3	Nanosensor 4	Nanosensor 5	IBM 1	IBM 2
Material	$Pr_2Fe_{17}B$	$Pr_2Fe_{17}B$	$Pr_2Fe_{17}B$	$SmCo_5$	Ferrit	$Nd_2Fe_{17}B$	$Pr_2Fe_{17}B$
$\mu_0 H_c$ [T]	0.7	0.7	0.7	3.2	0.2		0.7
$\mu_0 M_s$ [T]	0.986	0.986	0.986	1.05	0.4		0.986
K_1 [MJ/m³]	0.3	0.3	0.3	1.3	0.04		0.3
Mass [ng]	5	40	275	4	10	0.22	0.29
Volume [µmm³]	0.658	5.263	36.184	0.477	1.282	0.029	0.038
f_{i0} [Hz]	10734	10112	11020	10423	10523	2736	2801
f_0 [Hz]	8663.7	7321.92	3494.29	8881	8152	2158	2110
K_0 [n/m]	0.155	0.123	0.11	0.155	0.127	0.00014	0.00018
Q factor	94567	157580	102569	145335	110569	29064	28654
F_{min} [N/Hz$^{0.5}$]	$6.83 \cdot 10^{-16}$	$5.76 \cdot 10^{-16}$	$1.03 \cdot 10^{-15}$	$5.45 \cdot 10^{-16}$	$6.52 \cdot 10^{-16}$	$7.93 \cdot 10^{-17}$	$8.08 \cdot 10^{-17}$

Table 4.1: Magnetic properties. The table resumes and compares the magnetic saturation, the coercitiviy, the Curie temperature, the anisotropy constant and the maximal radius for a monodomain of different magnetic materials.

For small angles ($\sin(\theta) \approx \theta$ and $\cos(\theta) \approx 1$) we find:

$$\theta_m = \begin{cases} \frac{B}{B + \mu_0 M_z (D_\perp - D_\parallel)} \theta & (geometry - anisotropy) \\ \frac{B}{B + \frac{2K_1}{M_z}} \theta & (cristalline - anisotropy) \end{cases} \quad (4.10)$$

The torque $\tau(\theta, \theta_m)$ acting on the cantilever can be calculated using the first derivate of the total energy by minimizing the interaction with the magnetic field and replacing θ_m with the equation 4.10 we obtain:

$$\vec{\tau}(\theta) = -\frac{\partial E(\theta)}{\partial \theta} \quad (4.11)$$

Assuming small angles the torque [15],[16] can be described by the following relation:

$$\|\vec{\tau}(\theta)\| = \|\vec{\tau}(\theta_{mc}) + \vec{\tau}(\theta_{imc})\| = \begin{cases} -(\frac{B\mu_0 M_z^2 (D_\perp - D_\parallel) V}{B + \mu_0 M_z (D_\perp - D_\parallel)} + k_0 l^2)\theta \\ -(\frac{2BK_1 V}{B + \frac{2K_1}{M_z}} + k_0 l^2)\theta \end{cases} \quad (4.12)$$

The torque is then divided in two terms, the magnetic torque τ_{mc} that interacts with the magnetic field, and the torque independent of the magnetic field τ_{imc}. The torque independent of the magnetic field is mainly due to the elasticity of the cantilever. Consequently it can be neglected when studying tip magnetic field interaction.

4.2.2 Frequency shift and damping factor

The torque acting on the cantilever induces a frequency shift. This is measured by standard FM-detection: A lock-in amplifier, a frequency counter or a phase lock loop called PLL. The frequency shift and the damping factor can be calculated with the motion equation of a damped harmonic oscillator excited at constant amplitude.

$$m\ddot{x} + \Gamma \dot{x} + k_0 x + \frac{\tau(x)}{l^2} = A\sin(\omega t) \quad (4.13)$$

where m is the mass, Γ is the damping factor, k_0 the spring constant, l the length of the cantilever and A the exciting amplitude. From the equation 4.13, the mechanical resonance frequency is

$$\omega = \sqrt{\frac{k_0}{m} + \frac{\Delta k}{m}} = \sqrt{\frac{\tau_{imc}}{ml^2} + \frac{\tau_{mc}}{ml^2} - \left(\frac{\Gamma}{2m}\right)^2} \qquad (4.14)$$

where the change of the spring constant is Δk. The frequency shift can then be calculated with the Taylor approximation:

$$\Delta\omega = \omega - \omega_0 = \omega_0 \left(\sqrt{1 + \frac{\frac{\tau_{mc}}{ml^2} - \left(\frac{\Gamma}{2m}\right)^2}{\omega_0^2}} - 1 \right) \cong \omega_0 \frac{1}{2} \frac{\frac{\tau_{mc}}{ml^2} - \left(\frac{\Gamma}{2m}\right)^2}{\omega_0^2} \qquad (4.15)$$

Finally it is possible to compact the relation to

$$\frac{\Delta\omega}{\omega_0} = \frac{1}{2} \frac{\Delta k}{k_0} \qquad (4.16)$$

and calculate the relation between frequency shift and magnetic field [15],[16],[18].

$$\Delta\omega = \frac{1}{2} \frac{\omega_0}{k_0} \Delta k = \begin{matrix} \frac{\omega_0}{2k_0}\left(\frac{1}{l^2}\frac{B\mu_0 M_z^2(D_\perp - D_\parallel)V}{B+\mu_0 M_z(D_\perp - D_\parallel)} - \frac{\Gamma^2}{4m}\right) \\ \frac{\omega_0}{2k_0}\left(\frac{1}{l^2}\frac{2BK_1 V}{B+\frac{2K_1}{M_z}} - \frac{\Gamma^2}{4m}\right) \end{matrix} \qquad (4.17)$$

For large anisotropy constant ($K_1 >> BM_s$) the frequency shift should have a linear relation with the magnetic field. At higher magnetic field the frequency shift should tend asymptotically to a constant value ($\omega_w/k_0/l^2 K_1 V$). The first relation of the equation 4.17 depends on the geometry factor and it must be used for isotropic multi domain samples or soft materials. The second relation is directly correlated with the anisotropic constant of the magnetic domain. Consequently, this relation must be used for hard magnetic mono domains. Also, the damping factor Γ can induces an effect on the frequency shift.

4.3 The energy losses and the damping factor

In the previous section we have developed the relation that connects the magnetic field to the frequency shift. In this section we explain the damping factor that appear in the frequency shift relation. The damping factor comprises many independent mechanisms, which cause a lost of energy in the system.

One such loss is the thermoelastic relaxation [7]. The energy loss is caused by a delay of the elastic process between two points of the mechanical oscillator causing an irreversible process of energy loss. Experimental results show that the thermoelastic process is independent of the applied magnetic field.

An independent dissipation process is also measured, when a mechanical oscillator with magnetic tip is placed in a magnetic field or interacts with a magnetic sample. There are different mechanisms by which a variable magnetic field can couple to a material and loose the energy. The main loss mechanisms for magnetic materials in a magnetic field are hysteresis, conduction losses (eddy current), domain wall resonance, and electron spin resonance. The different mechanisms have diverse dependencies on material properties such as sample type, microstructure, frequency and temperature.

A third dissipation effect is due to the charge fluctuation between tip and sample. The variation of the electrostatic field induces a charge variation causing a dissipation of the energy.

The three effects are modeled with an independent spring and damper, where the spring is the phenomena in phase with the change and the damper is the out of phase loss process. Each independent process can be added in parallel as shown in figure 4.5 and 4.6.

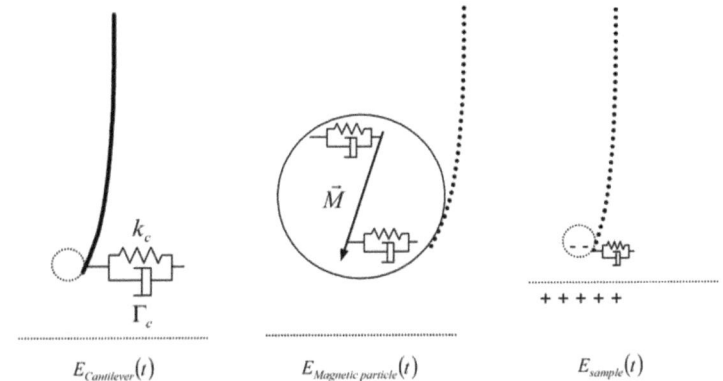

Figure 4.5: Cantilevers and dissipations. The sketch represents the different dampers acting on the mechanical lever. The first on the left represents the thermoelastic relaxation, the middle one the magnetoelastic relaxation and the third one the sample relaxation.

The total potential energy of the mechanical oscillator near a surface can be calculated by equation 4.4 in addition to a term defining the tip-sample interaction. The magnetic energy of the tip sample interaction can be calculated with the power spectral electrical density. The calculation of this effect is explicated in the chapter 6. The total potential energy and the losses are therefore:

$$E_t = E_{Cantilever} + E_{Magnetic\ particle} + E_{sample} \quad (4.18)$$

$$\Delta E_{Tot} = \Delta E_{Cantilever} + \Delta E_{Magnetic\ particle} + \Delta E_{Sample} \quad (4.19)$$

The quality factor is then calculated with the ratio between total energy and loss energy per cycle.

$$Q = \frac{2\pi E_{tot}}{\Delta E_{tot}} = \frac{1}{\frac{1}{Q_{Cantilever}} + \frac{1}{Q_{Magnetic\ particle}} + \frac{1}{Q_{Sample}}} \quad (4.20)$$

The quality factor contribution of each independent damping process is added in parallel. The following scheme represents the complete system with the noise excitation.

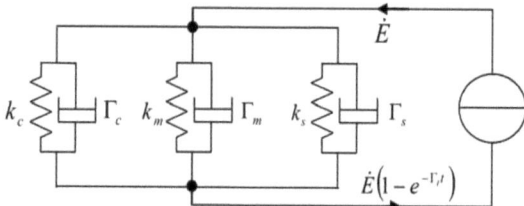

Figure 4.6: Energy dissipation processes. The different relaxation process are considered independent and can be added. All processes causing the dissipation energy are modeled with a spring assuming the elastic process and a damper causing the dissipation.

The three dissipation processes are the origin of the dephasing between excitation and detection. The dephase signal is calculated with the parameter Γ_t, which is the sum of the different dephasing processes.

4.3.1 Losses due to the oscillating magnetic field

Losses of energy are due to a change of phase between excitation and the detection. A central parameter, that is used in electrodynamics, and that gives the relation between the magnetic flux and the field is the permeability. The permeability, μ, of a material is defined by the relation $\mathbf{B}=\mu\mathbf{H}=\mu_0\mu_r\mathbf{H}=\mu_0(\mathbf{H}+\mathbf{M})$, where $\mathbf{B}$ is the flux density (T), $\mathbf{H}$ is the field intensity (A/m), $\mathbf{M}$ is the magnetization (A/m), μ_0 is the permeability of free space ($4\pi \times 10^{-7}$ H/m) and μ_r the relative permeability of the material. Losses which occur in a material because of the time varying magnetic field are included in the relative permeability term by writing μ_r, as a complex number, $\mu_r = \mu_{rr} - j\mu_{ri}$, where $j=\sqrt{-1}$ [19],[20].

The real term μ_{rr} describes the permeability at a fixed field without any losses. The imaginary term μ_{ri}, which describes the magnetic loss, arises from damping forces caused by internal friction during domain rotation and Bloch wall propagation.

Hysteresis losses: At low frequency this process dominates and dissipates as heat in a magnetic material as it generates **B-H** hysteresis loop. The energy loss per unit sample volume is $\Delta E = \int BdH$. This loss is controlled by factors that control low frequency permeability and coercivity such as porosity, grain size and impurity as well as the intrinsic properties. The energy used for turning the magnetization of θ_m degree is equal to $E_a = K_1 V \sin^2 \theta_m$.

Domain wall loss: At 100 kHz the small displacements of the pinned domain wall with the applied field introduce restoring forces. The wall has inertia and its movement is accompanied by energy dissipation [21],[22]. This process due to the frequency range can be ignored.

Eddy current losses: The eddy current depends with the frequency of the varying magnetic field and with the conductivity of the material. It is well known than when the skin depth $\delta = \sqrt{(1/(\sigma \pi f \mu))}$ is large compared to the sample size the influence of the eddy current on the magnetic field is entirely negligible [19],[20].

The mechanical resonance frequency of the cantilever is in the kHz range. At room temperature the best conductor has a penetration of half mm. This dimension is 1000 times bigger that the tip size dimension. At helium boiling temperature the skin depth of the majority of metal conductors is reduced by around 1000 times. Consequently, the use of conductor as Ni, Co or Fe in micron size range is affected by the eddy current. Rare earths magnets do not have a good conductivity and for this reason the eddy current can be neglected for such magnets.

The RF field excitation has a frequency of more than a Ghz. This high frequency is still not able to cause an eddy current loss in hard magnetic materials. This high frequency can cause a severe eddy current loss in the silicon cantilever. This is the reason why the RF coil is placed parallel to the lever surface.

The energy dissipation is $\Delta E = a\sigma f^2 B^2 r^2$, where a is a constant shape dimension, σ the conductivity, f the frequency, B the magnetic field and finally r the radius of the particle.

4.3.2 Tip-field interactions

The cantilever is placed in a homogenous static field. The vibration and the setup of the cantilever induces a small varying field, which generates a energy loss. The magnetic particle attached to the cantilever in a constant magnetic field is subjected to a variation field caused by the motion of the mechanical oscillator. It is possible to suppose the cantilever with the magnetic particle polarized in direction of the static magnetic field and perpendicular to this a small variation field. The representation of the model is show in the figure 4.7.

The magnetic field $\mathbf{H}_{ac}$ is calculated from the peak displacement x_{pk} and length l of the cantilever with the small angle assumption. The frequency ω is the frequency

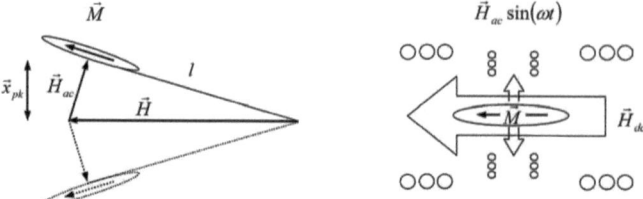

Figure 4.7: Tipped cantilevers in a static field. The magnetic field acting on the magnetic particle can be modeled with the sum of the axial and perpendicular compounds. The axial component for the low frequency sweeping can be shown as a quasistatic experiment. The perpendicular component has a frequency of oscillation of the mechanical resonator and for this reason has a strong effect on the frequency shift.

of the harmonic oscillator and **M** the magnetization of the particle.

The variation fields present inside the material induce rotation of the magnetic domains at kHz range. Since the induced motions are resisted by inertial, elastic and frictional forces the response is generally a function of the applied frequency [23]. In addition to the frequency dependence the response is a function of the temperature, the magnitude of magnetic field, the orientation and the magnetic domains [23],[24].

The figure 4.8 represents a hypothetic hysteresis loop caused by varying the static magnetic field. The red line shows the magnetization of the magnetic tip over the variation of the static magnetic field. Parallel to the field the process is elastic and dominated by the wall motions. Perpendicular to the magnetic field the motion is dominated by domains rotation and small hysteresis loops are generated, causing a loss of energy. Depending on the structure, it is possible to introduce a demagnetization factor that causes a reduction of the remaining magnetic field.

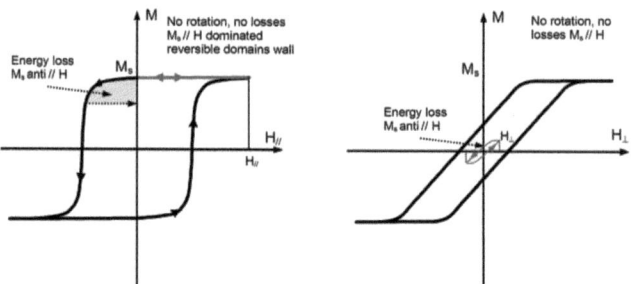

Figure 4.8: Hysteresis loops. The figure shows the hysteresis loop of a magnetic material. The axial components do not cause losses on the mechanical resonator, because the field is changed with a low frequency. Moreover in the red part of the left graph the domains are parallel to the field and are in lower potential. The right graph shows the real cause of the dissipation, by varying the field perpendicular to the magnetic field a small loop is engendered.

The magnetic field in the particle is $\mathbf{H}=\mathbf{H}_{//}+\mathbf{H}_{\perp}=\mathbf{H}\cos(\theta_m)+\mathbf{H}\cos(\theta_m)\sin(\omega t)$. In a continuous wave electron spin resonance signal the applied field is normally swept from a field $\mathbf{H}$ parallel to the particle magnetization to a field $\mathbf{H}$ antiparallel. This sweep causes a loss of energy marked in yellow on the left of the figure 4.8, but not detectable by the cantilever. In fact, the loss is not correlated with the vibration frequency of the mechanical oscillator and for this reason the Q factor is constant. This supposition is true only when:

Frequency sweep field $\ll$ Frequency cantilever

This relation demonstrated what many authors have measured [15],[16],[18], but never explained. In other words the energy loss is filtered by the mechanical resonator. The hysteresis loop is dependent on the magnetic oscillation frequency and is deformed as shown in the figure 4.9. The figure represents the real loop of $Fe_{10}Co_{90}$ (for the measured hysteresis loops please refer to [25]).

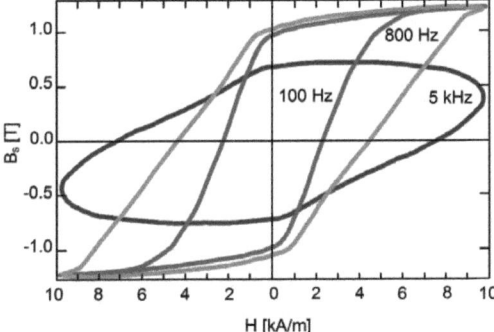

Figure 4.9: Hysteresis loop of FeCo. The graph represents the hysteresis loop of $Fe_{10}Co_{90}$ as a function of the frequency measured by Giri and All [25]. On the origin of the graph the change of the magnetization begins to be more difficult, because the magnetization can not follow the fast change of the induction field. For small loops the energy dissipated is reduced with increasing frequency.

The more the frequency increases, the less the magnetic domain due to its inertia follows the magnetic field. The energy loss, that is induced by the complete hysteresis loop, increases when the frequency increases. But the minor loop has a tiny decrease caused by the change of the permeability slope as a function of the field. The anisotropy slightly increases when the excitation magnetic field frequency increases. This effect is due to the inertia of the magnetic domains.

The total energy loss per cycle corresponds of the area of the minor loop. At frequency lower than 1 kHz the domains follow the induced magnetic field. At higher frequency (> 1-2kHz) the domains due to their inertia hardly follow the magnetic field. So more energy is needed for turning the magnetization and completing the loop. At smaller magnetic field amplitudes and at higher frequencies the domains do not follow the field variation. The process consequently dissipates less energy.

4.3.3 Magnetic interaction losses

The damping measured through the Q factor provides an important parameter of the dissipation particle material and the imaginary part of the magnetic permeability. In MFM, one usually measures the frictional constant, which is related to the imaginary magnetic permeability. The energy magnetic loss is induced by the time varying magnetic field, which produces an amount of energy dissipation during each period. The total energy loss is extrapolated from the linear dispersive media theory of losses [19],[20]. The theory shows that at a given instant in time and space, the rate of heat generated per unit volume caused by magnetic losses is given by:

$$P_{losses} = \omega \mu_0 \mu_i H_{ac}^2 \tag{4.21}$$

In a complete cantilever oscillation period T, the energy dissipated by a particle with volume V is therefore:

$$\Delta E = P_{losses} T = 2\pi \mu_0 \mu_i V H_{ac}^2 \tag{4.22}$$

The alternative magnetic field excitation can contribute of the generation of phonon and dissipate the energy by the relaxation processes though the magnetic material. This dissipated energy can be connected with the dissipated energy measured by the Q factor and the amplitude oscillation. The energy loss for a cantilever oscillating with amplitude x_{pk} is given by

$$\Delta E = \frac{2\pi E}{Q} = \frac{\pi k_0 \omega^2 x_{pk}^2}{Q \omega_0^2} = \pi \omega x_{pk}^2 \Gamma \tag{4.23}$$

where $\Gamma = k_0 \omega / \omega_0^2 / Q$ represents the total friction [11]. The imaginary part of the magnetic permeability can be consequently determined by substituting the variable field H_{ac} with Hx_{pk}/l and by comparing the equation 4.21, 4.22. The total friction and the imaginary magnetic permeability are two parameters which determinate the amount of dissipated energy. The imaginary permeability is mostly used in the high frequency magnetic field instruments for calculating the energy dissipation. The MFM is a strong sensitive instrument, which can measure small variation of this parameter.

Chapter 5

Experiments: Tip-field interactions

In the previous chapter we calculate the interaction between the magnetic field and microsized magnetic particles. At small magnetic fields, a linear relation connects the magnetic field, which produces a torque on the mechanical beam, and the frequency shift. The friction and the imaginary magnetic permeability have been calculated as well. Thus, we conducted a series of experiments, in order to verify the frequency shift relation calculated in the previous chapter and to understand the severe energy losses while exposing the cantilever to the magnetic field.

5.1 Tip materials and setup

In this study we glued various grain sizes and magnetic materials on 5 Nanosensors[1] cantilevers with a mechanical resonance frequency of 10 kHz and on two ultrasoft IBM[2] cantilevers with a mechanical resonance frequency of 2.7 kHz. The magnetic materials and cantilever characteristics are shown in the table 5.1, before and after gluing the tip.

In order to attach the magnetic particle at the end of the ultrasoft cantilever a really minuscule quantity of optical glue[3] is placed at the extremity of the cantilever using an optical microscope and a home build micromanipulator. Small magnetic particles were placed on an AlO_2 substrate and scratched against an AlO_2 substrate with the purpose of reducing the dimension of the grains. After choose the particle, the permanent magnet is mounted at the end of a cantilever.

---

[1]NANOSENSORS^{TM}, Rue Jaquet-Droz 1, CH-2007 Neuchatel, Switzerland http://www.nanosensors.ch.

[2]IBM Zurich, Säumerstrasse 4, CH-8803 Rüschlikon, Switzerland, http://www.IBM.ch.

[3]Norland optical adhesive 65, Norland Products, P.O. Box 637, 2540 Route 130, Suite 100, Cranbury, NJ 08512, http://www.norlandprod.com.

Cantilever	Nanosensor 1	Nanosensor 2	Nanosensor 3	Nanosensor 4	Nanosensor 5	IBM 1	IBM 2
Material	$Pr_2Fe_{17}B$	$Pr_2Fe_{17}B$	$Pr_2Fe_{17}B$	$SmCo_5$	Ferrit	$Nd_2Fe_{17}B$	$Pr_2Fe_{17}B$
$\mu_0 H_c$ [T]	0.7	0.7	0.7	3.2	0.2		0.7
$\mu_0 M_s$ [T]	0.986	0.986	0.986	1.05	0.4		0.986
K_1 [MJ/m³]	0.3	0.3	0.3	1.3	0.04		0.3
Mass [ng]	5	40	275	4	10	0.22	0.29
Volume [μmm³]	0.658	5.263	36.184	0.477	1.282	0.029	0.038
f_e [Hz]	10734	10112	11020	10423	10523	2736	2801
f_0 [Hz]	8663.7	7321.92	3494.29	8881	8152	2158	2110
K_0 [n/m]	0.155	0.123	0.11	0.155	0.127	0.00014	0.00018
Q factor	94567	157580	102569	145335	110569	29064	28654
F_{min} [N/Hz$^{0.5}$]	$6.83 \cdot 10^{-16}$	$5.76 \cdot 10^{-16}$	$1.03 \cdot 10^{-15}$	$5.45 \cdot 10^{-16}$	$6.52 \cdot 10^{-16}$	$7.93 \cdot 10^{-17}$	$8.08 \cdot 10^{-17}$

Table 5.1: **Cantilevers and tip characteristics.** The table shows the cantilevers on which a magnetic particle material has been glued. The materials are $Pr_2Fe_{17}B$, $SmCo_5$, Ferrit and $Nd_2Fe_{17}B$.

The particle is localized and aligned with the magnetic field (it is strongly recommended not to place the magnet behind the magnetic particles, whereupon the particles become fixed and impossible to remove from the surface). After the particle is captured by the glue the cantilever is exposed to UV rays for few hours to harden the adhesive. Finally, all cantilevers are photographed using the SEM microscope and placed in a plastic box filled of Argon gas with the aim of reducing the oxidation of the magnetic particle.

The mass of the particle attached on the cantilever is determined using the frequency shift. This is determined by the difference between the first eigenfrequency of the cantilever before attaching the magnetic particle to it, and after. The mass is then calculated. For cross validation the mass is determined by reconstructing the volume from the Scanning Electron Microscope (SEM) pictures represented in the table 5.2 multiplied with the mass density.

To demonstrate and measure the magnetic dissipation a dynamic mode cantilever was measured in vacuum at 10^{-6} mbar and at room temperature between the poles of an electromagnet (Bruker maximal magnetic field 0.5 T). The basic experiment was to measure the cantilever resonance frequency and the damping as a function of the static magnetic field. Cantilever frequency was measured with a Labview program with the FFT of a signal acquired for 30 sec with a precision of 0.1 Hz, while damping was typically measured by the cantilever ring-down time after abruptly turning off the piezoelectric drive signal.

The measurement was repeated after the cantilever amplitude reached the steady state condition for a fixed magnetic field. The lock-in is adjusted for the maximal sensitivity and the local oscillator, exciting the piezo, turned off. The measurement is repeated 30 times and then the field changed to the next amplitude. The scheme of the equipment used for this measurement is sketched in figure 5.1.

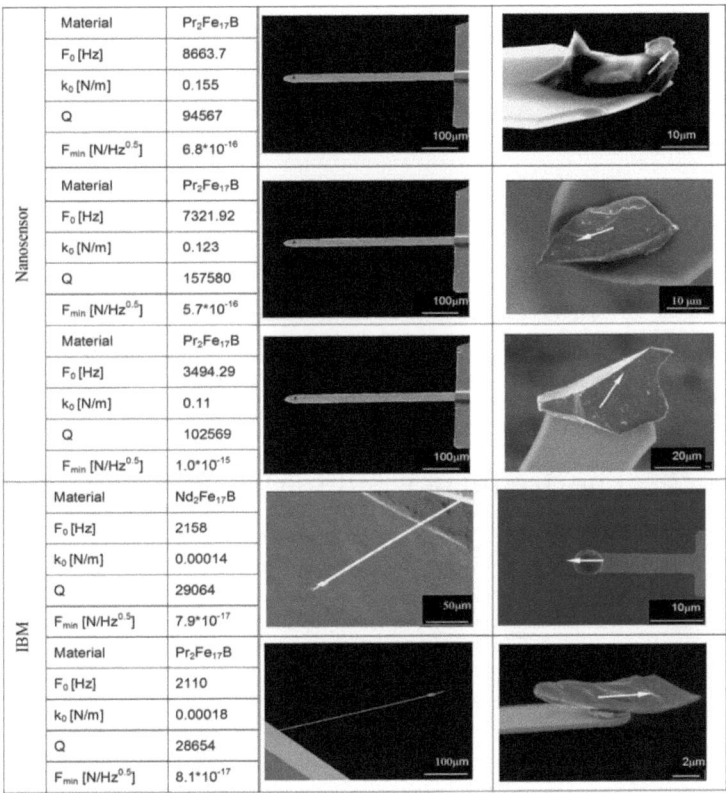

Table 5.2: Tipped cantilevers. The table shows the cantilevers measured and their parameters, such as the resonance frequency, the spring constant, the Q factor and the minimal detection force at room temperature.

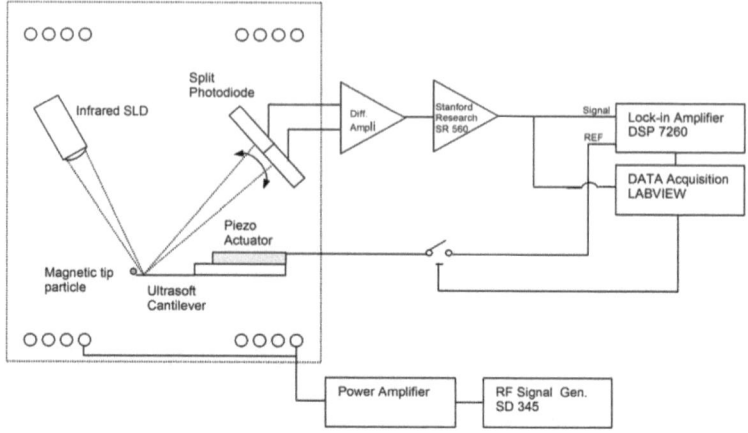

Figure 5.1: Tip-field interaction setup. The schematic diagram sketches the circuit used to measure the eigenfrequency and the Q factor of the cantilever with the magnetic particle. The Labview program measures the eigenfrequency of the cantilever by FFT. The lock-in is then set for the maximal sensitivity and when the oscillation of the cantilever is stable the local oscillator is turned off. The signal is measured for the decay time and repeated for different magnetic fields.

5.2 The frequency shift as a function of the magnetic field

The frequency shift is measured for the cantilevers mentioned in the previous section. The frequency shift of the tipped cantilevers changes linearly with the magnetic variations. This tendency is experimentally demonstrated with various hard magnetic materials. When the anisotropy constant is larger than the product between the magnetic field and the particle magnetization: $K_1 >> BM_s/2$ or $(D_\perp - D_{//})\mu_0 M_s^2/2 >> BM_s/2$, the slope can be fitted with the following linear equation:

$$\Delta\omega = \left(\frac{1}{2}\frac{\omega_0 V M_s}{k_0 l^2}\right) B - \frac{1}{8}\frac{\omega_0^3}{k_0^2 l^2}\gamma^2 \qquad (5.1)$$

The frequency shift depends linearly with the particle volume and its magnetization. The tipped cantilever with a hard magnetic particle is a relative magnetometer with a strong linearity for constant temperature. Since the temperature of the cantilever is not controlled the variation perturbs the measurement and causes some oscillations.

In case the anisotropy factor is equal or smaller than the product between the magnetic field and the particle magnetization the slope changes and reduces to 0. The ferrit particle shows this effect, since it has a change at 0.2 T. In this case equation 4.17 should be used for fitting the curve. The tabulated magnetic saturation of ferrit is at 0.4 T. So, the anisotropy constant is: $K_1 = BM_s/2 = 0.04$ MJ/m^3. The inflection can be used consequently to calculate the anisotropy constant of each magnetic particle. We report the results in figure 5.2.

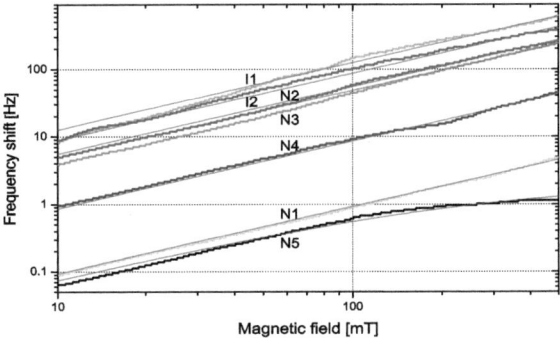

Figure 5.2: Frequency shift vs magnetic field. The graph shows the dependence of the frequency shift as a function the static magnetic field applied. The labelling is the following: I stay for IBM and N stay for Nanosensors; for the numbers see table 5.1. All materials have a linear tendency and are fitted with equation 5.1. Only the ferrit N5 shows saturation and is fitted with equation 4.17. The change of temperature causes the perturbations.

Soft materials have small anisotropy and consequently their magnetization is changed with the oscillations of the cantilever. This change causes not only a rotation of the magnetization but also a strong fluctuation field. This introduces a broadening of the change of the magnetic field interacting with the spins and consequently an increase of the relaxation time rate spins. The magnetization of the hard material is also turned but with a smaller amplitude. For this reason, we focus our study on hard magnetic materials.

5.3 The quality factor measurement as a function of the magnetic field

In the previous chapter the quality factor is connected with the friction and the imaginary part of the permeability. Consequently, the Q factor is the parameter for measuring the dissipation. The quality factor of the mechanical resonator is measured using the ring down measurement technique. Each measurement is repeated 30 times for a constant magnetic field and averaged. The data are represented in graph 5.3.

The Q factor of the cantilevers is fitted with two anelastic processes, one caused by the thermo elastic relaxation, and one caused by the interaction tip-magnetic

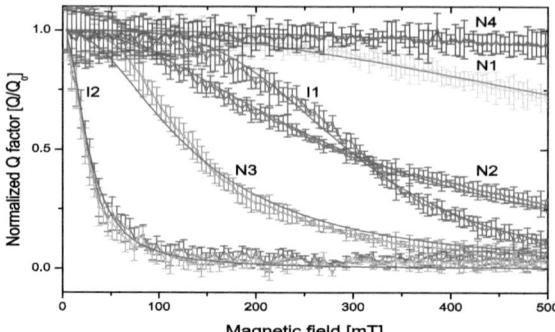

Figure 5.3: The graph shows the normalized Q factor of the different mechanical levers. All trends are fitted using the equation 4.22. The change in the Q factor is directly proportional to the magnetic material glued, and inversely proportional to the anisotropy constant. The labeling is the following: I stay for IBM and N stay for Nanosensors; for numbers see table 5.1.

field. The thermoelastic relaxation is not correlated with the magnetic field. The Q factor change is directly proportional with to the volume of material glued and inversely proportional with to the anisotropy. The Q factor of the mechanical lever N4 does not have a strong change, because $SmCo_5$ has an anisotropy value 5 times larger than the anisotropy of the $Pr_2Fe_{17}B$ material glued at the cantilever N1, N2, N3 and I2.

The magnetic friction losses for each magnetic material are extrapolated from the Q factor measurements. The magnetic friction is then divided by the volume of the magnetic particle. The density friction can be plotted and compared. The graph 5.4 represents the magnetic losses per nm^3 of magnetic material. The frictional loss of the isotropic $Pr_2Fe_{17}B$ material is represented by four curves with an offset in between.

This curious effect can be explained by the frequency oscillation of the mechanical lever. In fact as explained in the previous chapter the anisotropy constant is correlated with the frequency oscillation. In the case of minor loop hysteresis the energy losses is smaller, because the magnetization of the particle has inertia to turn.

The $SmCo_5$ has an anisotropy constant of 1.3 MJ/m^3, about 12 times lower than a monodomain. The large anisotropy constant may explain the constant behavior in the range of 0.5 [T]. In the single spin experiment, Rugar [4] has used a submicrometer magnetic tip of $SmCo_5$ for its incomparable anisotropy.

The vitreous $Nd_2Fe_{17}B$ spherical material has a curious strong magnetic friction that is difficult to explain. It may be attributed to the fact that the oxidation has dramatically decreased the anisotropy constant.

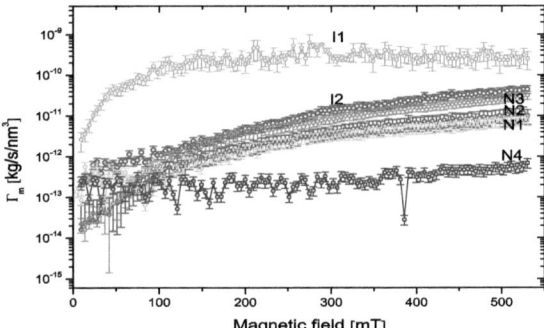

Figure 5.4: The graph represents the magnetic friction per nm^3 of material as a function of the magnetic field. The dissipation is correlated with the frequency. The higher the frequency the lower the dissipation. $SmCO_5$ has highest anisotropy and consequently it has the lowest dissipation. The labeling is the following: I stay for IBM and N stay for Nanosensors; for numbers see table 5.1.

The magnetic frictional losses could be further decreased by reducing the particle to a monodomain dimension of 0.8 μm. In this case all domain wall losses would disappear and the losses would be caused only by the hysteresis loop of the magnetic domain. The anisotropy is increased to 17 MJ/m^3 at room temperature, to 24 MJ/m^3 at nitrogen boiling temperature and finally to double the value at helium boiling temperature.

At the actual knowledged, we know that the domains are hardly turned at higher frequency. In fact increasing the frequency causes a decreasing of the magnetic friction. The data are plotted in figure 5.5. This behavior is caused by the minor loop, where the field due of the frequency change is not able to turn the domains.

The friction for volume samples has a dependency with the frequency oscillation. Lower frequencies dramatically increase the dissipation due the hysteresis effect. For frequencies higher than 10 kHz the friction is independent of the frequency [15]. The wall resonance, the eddy current and the electron spin resonance will increase the friction at higher frequency.

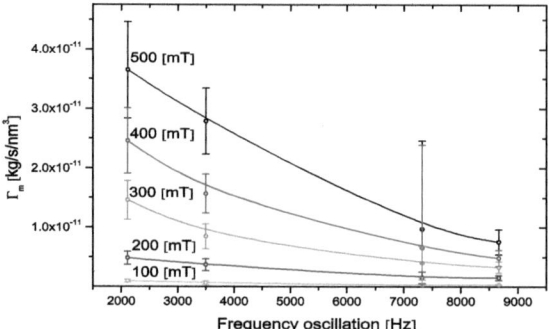

Figure 5.5: Magnetic friction vs frequency oscillation. The graph shows the trend of the magnetic dissipation per nm^3 as a function of the frequency of the different Nanosensors cantilevers. The trend increases at higher magnetic field.

5.4 The force sensitivity as a function of the magnetic field

The force sensitivity is strongly affected by the tip magnetic field interaction. The choice of the right magnetic material is consequently fundamental to maintain a high sensitivity in order to measure single electron spin. The theoretical minimum measurable detecting force is given by the following equation:

$$\frac{F_{\min}}{\sqrt{\Delta f}} = \sqrt{\frac{4k_0 k_B T}{w_0 Q}} = \sqrt{4\Gamma_m m_p k_B T} \quad (5.2)$$

Experimentally, it is found that the frequency shift and the spring constant hardness do not have as strong an influence on the force sensitivity as the Q factor. The force sensitivity reported in the figure 5.6 or figure 5.7 is calculated as a function of one parameter only: either the Q factor or the spring constant. In this way, the sensitivity change can be compared. The magnetic damping loss is the major factor that reduces the force sensitivity in a static field below 0.5 T.

A maximal sensitivity of 7.5x10^{-17} N/$\sqrt{Hz}$ at room temperature for the IBM cantilever is extrapolated. This force sensitivity can be increased to 9x10^{-18} N/$\sqrt{Hz}$ by reducing the temperature to the helium boiling temperature. This sensitivity is an under estimation, because the anisotropy constant increases while the temperature decrease. The same experiment should be performed at helium boiling temperature. Moreover, the sensitivity can be increased further by annealing the cantilever. In this case the grain glued at the end of the cantilever must to be a monodomain in order to decrease the correlated demagnetization factor. The $Ne_2Fe_{17}B$ material

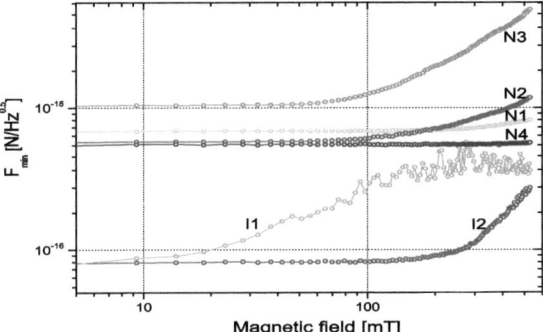

Figure 5.6: Sensitivity as a function of the magnetic field and Q factor. The graph shows the force sensitivity vs the magnetic field as a function of the Q factor. The spring constant and the frequency shift are supposed constant.

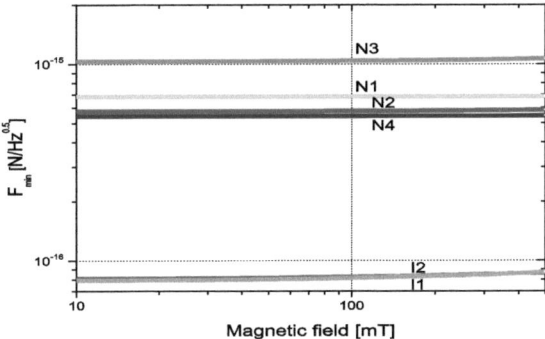

Figure 5.7: Sensitivity as a function of the magnetic field and frequency shift. The graph shows the force sensitivity vs the magnetic field as a function of the frequency shift and the spring constant. The Q factor is supposed constant.

shows a curious loss of force sensitivity for fields of more than 10 [mT]. The $Pr_2Fe_{17}B$ and the $SmCo_5$ have a much better behavior and the force sensitivity holds for more than 100 [mT] at room temperature.

The anisotropy plays a central role in the friction process as a function of the magnetic field. In fact the sensitivity and the magnetic field range are directly correlated with the magnetic anisotropy of the particle attached to the cantilever. Low temperature and hard magnetic materials increase the magnetic anisotropy constant, which decreases the magnetic friction.

Chapter 6

Tip-sample interactions and damping losses

Friction and frequency are measured while approaching the surface with a tipped cantilever. The vertical cantilever is sensitive to forces and force gradient interactions in a different manner than a standard horizontal cantilever. In this chapter we will expound on the forces and force gradient interacting with a vertical cantilever. The tip-sample interaction has to be understood in order to optimize the detection sensitivity.

6.1 Tip-sample interactions

In this section, we report the tip sample interaction measured between a Sm_2Co_{17} tip (5 μm long and 4 μm wide) mounted on an IBM cantilever and an irradiated quartz gold plated sample(Suprasil 300) at room temperature. Diamagnetic forces, electrostatic forces interact between the sample and the magnetic tip. The force gradient and the vertical force induce a change in the stiffness of the cantilever and consequently a frequency shift.

The frequency measured with this method is different from what we expect from standard measurement. In fact the strong magnetic gradient and the non compensated charges should generate a repulsive force causing drop in the stiffness.

6.1.1 Frequency as a function of the distance: theory and measurement

A permanent magnet of Sm_2Co_{17} is manually mounted on a IBM cantilever using a microscope and a micromanipulator. The dimensions of the grain mounted, as illustrated in figure 6.1, is 5 μm long, 4 μm wide and 3 μm height. The experiment is performed without any external magnetic field.

The tip and the sample are grounded to the same potential. An attractive electrostatic force is present between tip and sample, since the charges are not compensated by the difference of potential between tip and sample. The magnet-tipped cantilever is approached to the surface perpendicularly.

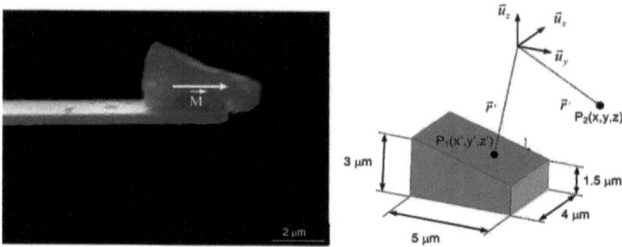

Figure 6.1: IBM cantilever with magnetic tip. The magnetic material is Sm_2Co_{17} with a coercivity field of 496 mT. On the right side the magnetic tip is modeled as a prism.

The fused silica sample with paramagnetic defects (Suprasil 300), is gold plated. This allows us to compensate the contact potential between sample and tip. The sample is a approached to the cantilever with the stack motor in the long range (1-20 μm) and subsequently with the piezo tube in the short range (less than 1 μm). All measurements are performed at room temperature and under UHV.

A frequency shift is detected while approaching the surface with the tipped cantilever. The frequency shift of the approaching cantilever in a vertical configuration has an opposite trend of what we expect. In fact, an attractive force gradient should decrease the frequency shift and a repulsive one increase the frequency shift. So, in order to understand which effect is really causing this frequency shift, we need to identify the interaction that can cause a frequency shift on a vertical cantilever.

A vertical cantilever is sensitive to vertical forces (parallel to the cantilever axis) and a transverse gradient (perpendicular to the cantilever axis), as represented in figure 6.2. We can assume that an attractive force causes a tension on the cantilever, which causes an increasing of the frequency shift. Conversely, a repulsive force causes a decrease in the frequency shift. In fact, this will happen even in the absence of a force gradient. A uniform force can cause an increase just as a uniform gravitational force will causes a pendulum to have a frequency that is proportional to the gravity.

The lateral force gradient induces hardness or softness to the stiffness of the mechanical lever, so a positive gradient will cause an increase in the frequency and a negative gradient a decrease in the frequency. Just like what appends in a standard parallel configuration.

These two phenomena induce the total frequency shift, caused by different kind of long-range interactions such as the electrostatic interaction, the diamagnetic interaction, and finally the Van der Waals interaction. In order to understand which phenomena mainly cause the vertical interaction or even which phenomena simultaneously interact we have evaluated the different phenomena independently.

Suppose that the gradient lateral force and vertical force are independent. Then the total spring constant is the sum of the force gradient component and the axial force component.

$$\vec{k}_{total} = \vec{k}_{cantilever} + \vec{k}_G + \vec{k}_F \tag{6.1}$$

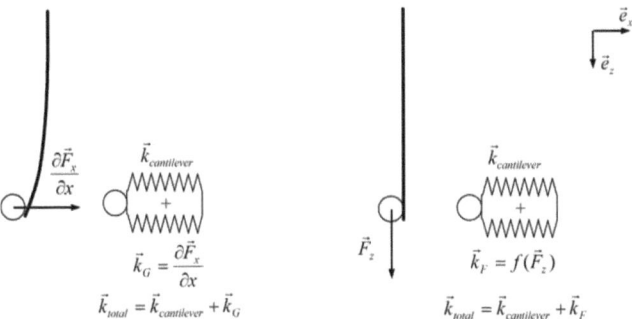

Figure 6.2: Force gradient and force interaction. The figure shows the lateral gradient interaction and the vertical force interaction for a positive gradient and axial tensile force the frequency shift is positive.

In a homogenous surface, the gradient component should be zero. However in most of the cases, the cantilever and the sample are slightly tilted, and the surface is inhomogeneous, which creates a lateral force gradient. We can measure the lateral force gradients by scanning the sample and measuring the frequency as a function of sample position. We will find regions of both positive and negative frequency shift. Moreover, the particle is glued on a side of the cantilever, and in addition to an attraction force a tensile surface stress and a momentum are created causing a bending of the cantilever.

6.1.2 The axial force and the frequency shift

Many publications clearly explain the relation between force gradient and frequency shift where the relation is proportional to the square of the spring constant over the effective mass [12]. However, the relation between the vertical force and frequency shift is uncommon, because in the standard MFM experiments the cantilever approaches the surface horizontally. The frequency shift in a horizontal configuration shift is not affected by vertical forces.

In our case, the largeness of the tip and the setup makes the axial force important. This results in a modification of shape and frequency [14]. The motion equation of the mechanical resonator including the axial force **N** which is time independent is:

$$EI\frac{\partial^4 v}{\partial x^4} + N\frac{\partial^2 v}{\partial x^2} + \bar{m}\frac{\partial^2 v}{\partial t^2} = 0 \qquad (6.2)$$

The equation can be analytically solved using the separation of variables technique and as a function of the axial force, the frequency shift of the various eigenmodes calculated with the following relation (the complete resolution is explained in the appendix A.1).

$$\frac{\varepsilon \sin(\varepsilon l) + \delta \sinh(\delta l)}{\varepsilon^2 \cos(\varepsilon l) + \delta^2 \cosh(\delta l)} = -\frac{\delta^2 \cos(\delta l) + \varepsilon^2 \cosh(\varepsilon l)}{\varepsilon \delta^2 \sin(\delta l) - \delta \varepsilon^2 \sinh(\varepsilon l)} \quad (6.3)$$

where l is the length of the cantilever, and ε and δ are defined as

$$\varepsilon = \sqrt{\left(a^2 + \frac{g^4}{4}\right)^{\frac{1}{2}} - \frac{g^2}{2}} \qquad \delta = \sqrt{\left(a^2 + \frac{g^4}{4}\right)^{\frac{1}{2}} + \frac{g^2}{2}} \quad (6.4)$$

and

$$a^4 = \frac{\bar{m}\omega^2}{EI} \qquad g^2 = \frac{N}{EI} \quad (6.5)$$

where $\bar{m}$ is the mass pro length, ω the eigenfrequency, I the moment section and E the Young's modulus of the cantilever (please refer to annexe A.1). The calculation results are represented in figure 6.3, in figure 6.4 and in figure 6.5 for a static attractive and repulsive force (the frequency shift is presented with the absolute value).

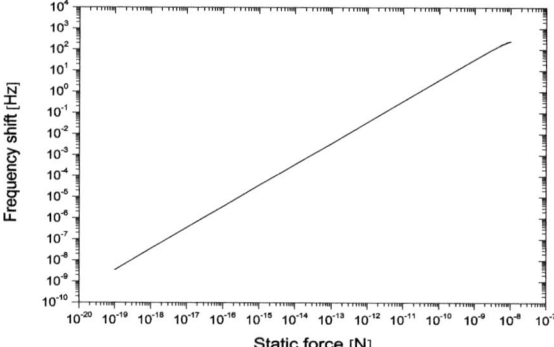

Figure 6.3: Axial tensile force. The graph represents the relation between pure axial force and frequency shift for tensile force acting on a IBM cantilever with a spring constant of k=0.175 mN/m.

It can be shown that when the acting force is tensile, the natural frequency increases and when is compressive the frequency of the transverse vibration decreases [26]. The frequency changes due to tensile stress have been reported [27]. The axial stress exhibits the change of the stiffness. The graph 6.3 shows that for static forces larger than 10^{-9} N the frequency shift has a non-symmetrical behavior.

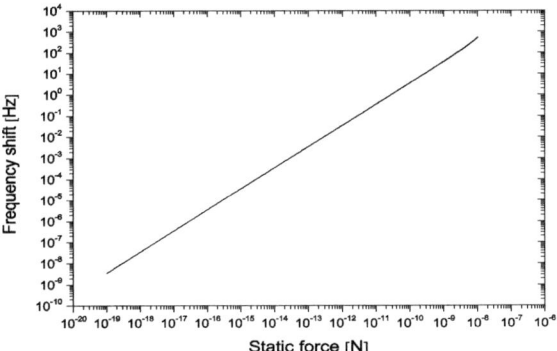

Figure 6.4: Axial compressive force. The graph represents the relation between pure compressive force and frequency shift acting on a IBM cantilever with a spring constant of k=0.175 mN/m.

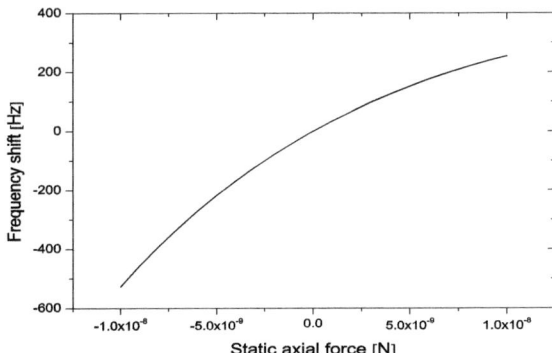

Figure 6.5: Axial tensile and compressive force. The graphs represents the relation between pure axial force and frequency shift for tensile and compressive forces acting on a IBM cantilever with a spring constant of k=0.175 mN/m. The relations between frequency shift and tensile or compressive forces are not symmetric. A tensile force induces an increase in the frequency shift and inversely a compressive force a decrease of frequency. Frequency shift of Hz range is caused by an axial force in the order of 10^{-11} N.

6.1.3 The electrostatic force

The electrostatic force occurs on metal and semiconductor surfaces if a voltage is applied or if the potentials of the tip and sample materials are not the same. The electrostatic force, which is linear distance dependent, is calculated by the difference of electrostatic potential between sample and tip. The force has a quadratic behavior as a function of the the difference of potential ΔU and consequently a minimum, called contact potential [28].

The electrostatic force between a sphere of radius R and an infinite grounded plane can be analytically determined by calculating the sphere-plane capacitance given by the following relation [29]:

$$C(d) = 2\pi\varepsilon_0 R \int_0^\pi \frac{\sin^2(\theta)}{\theta\left(\frac{d}{R} + 1 - \cos(\theta)\right)} d\theta \qquad (6.6)$$

$$\cong 2\pi\varepsilon_0 R \ln(\frac{d+R}{d}) \qquad (6.7)$$

The force can be calculated from the derivative of the capacitance and the result is:

$$F_c = -\frac{1}{2}\frac{\partial C}{\partial d} V^2 \qquad (6.8)$$

$$= 2\pi\varepsilon_0 \left(\Delta V^2 + V_0^2\right) \sum_{n=1}^\infty \frac{\coth(a\cosh(1+\frac{d}{R})) - n\coth(na\cosh(1+\frac{d}{R}))}{\sinh(na\cosh(1+\frac{d}{R}))} \qquad (6.9)$$

$$\cong \pi\varepsilon_0 \left(\Delta V^2 + V_0^2\right) \frac{R}{d} \qquad (6.10)$$

where ε_0 is the electric permeability, ΔV the difference of potential between tip and surface, $V_0 \sim 0.2$ V characterizes the electrostatic force at zero bias [11], R the radius of the sphere and d the distance between sphere and plane.

The contact potential between tip and sample is at -1.1 V, when the sample is grounded. At micron distance this potential produces an electrostatic force of 10^{-11} N as represented in graph 6.6. The graph represents the electrostatic tip-sample force for potential between 0.1 V to 1.8 V. The force as a function of the potential has a quadratic behavior and consequently can change in only 1 V of potential applied of 100 times.

As mentioned before, the frequency shift measured and reported in figure 6.7 shows the opposite behavior of what is expected by changing the potential. The contact potential is measured around the minimum frequency. In a standard configuration, where the cantilever is positioned parallel to the surfaces the contact potential is measured around the maximum frequency.

We know for experience [29],[30] that the electrostatic force part of the total interaction for a metallic tip-sample system is always attractive. The horizontal cantilever is sensitive to the force gradient, and due to the attractive force gradient the frequency shift is reduced. The vertical cantilever is sensitive to the vertical force, and due to the attractive behavior the frequency shift is increased.

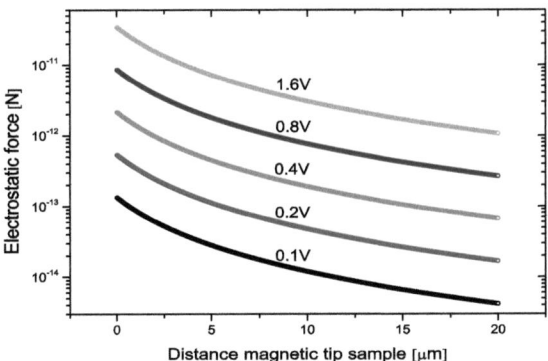

Figure 6.6: Electrostatic force as a function of the distance. The graph represents the electrostatic force generated by a sphere of radius 2.02 μm near an infinity grounded plate. The potential between the two objects are 1.1 V plus the 0.2 V of in homogeneity charges. The force generated is 10^{-11} N.

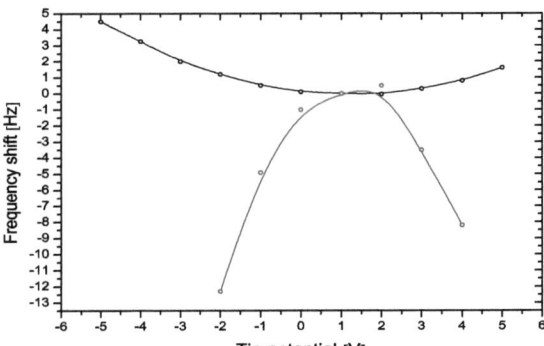

Figure 6.7: Frequency shift as a function of the potential. The graph represents the frequency shift as a function of the potential applied to the tip (sample grounded) for a vertical cantilever (line to the top) and horizontal cantilever (line on the bottom). These two measures are made from two different IBM cantilevers with same tip material with different size. The sample material is different one is quartz (top graph) and one is pyrolic graphite (bottom graph).

In order to understand the difference between the two configurations, we have compared the horizontally cantilever with high oriented pyrolitic graphite (HOPG) sample and the vertical cantilever with the gamma irradiated ^{60}C rays quartz(Supracil 300) sample. Even if the two configurations have different materials and different sample tip distances, we can extrapolate the trend of the frequency shift as a function of the the electrostatic potential applied.

The horizontal cantilever has the disadvantage that when the potential is increased, it is strongly attracted and finally it sticks to the sample surface. This force, which is purely an attractive force, does not cause any frequency shift change. Even if the vertical configuration is subjected to the same vertical force, the cantilever is stretched and can not attach to the sample surface. On the other hand, the frequency shift of the vertical cantilever dependents on the vertical force.

Consequently, the different trend reported in the graph 6.5 is induced from different interactions. One is caused by the attractive force acting on the vertical cantilever, and one by the force gradient that cause a decrease of the frequency shift.

6.1.4 The diamagnetic force

The permanent magnet tip mounted on the mechanical lever interacts strongly when it is positioned near the diamagnetic sample. The diamagnetic force is generated by the strong gradient produced by the tip, around 0.5 T/μm and the magnetized sample. The diamagnetic effect produces a repulsive force, which can be calculated knowing the susceptibility of the sample. The susceptibility of the gamma irradiated ^{60}C rays quartz quartz is $\chi = -4.93 \times 10^{-7}$ and the susceptibility of HOPG is $\chi_\perp = 450 \times 10^{-6}$.

The magnetic field $\mathbf{H}(\mathbf{r})$ generates by the permanent magnet is calculated with the following integral:

$$\vec{H}(\vec{r}) = \frac{1}{4\pi} \oint_S \sigma(\vec{r}\,') \frac{\vec{r} - \vec{r}\,'}{\|\vec{r} - \vec{r}\,'\|^3} dS \qquad (6.11)$$

where the magnetic charge density $\sigma(r')$ satisfies the following equation:

$$\sigma(\vec{r}\,') = \vec{n}(\vec{r}\,') \cdot \frac{\vec{B}(\vec{r}\,')}{\mu_0} \qquad (6.12)$$

The integral represents the magnetostatic equivalent of the Coulomb-law in electrostatics for the computation of the electric field.

The integral is calculated over the entire surface of the tip object of figure 6.1. The magnetic field generated by the magnetic tip is represented in the figures 6.8, 6.9 and 6.10 as a function of the distance. The first two graphs represent the field distribution in the perpendicular plane with respect to the distance, and the third represents the magnetic field gradient. At smaller distance (distance smaller than the particle size) the gradient field edges become important and strong variations in the magnetic field are induced.

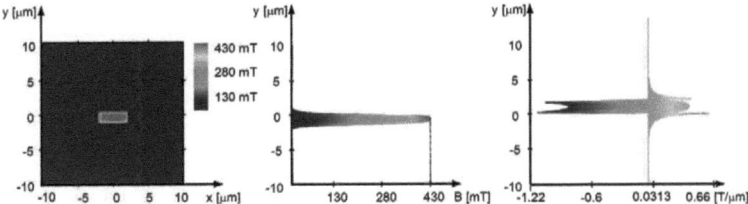

Figure 6.8: Magnetic field and gradient at 100 nm. At smaller distances than the tip size the field and the field gradient change rapidly for the edge effects. At 100nm the maximal field generated is 430 mT and the gradient field is 1 T/ μm.

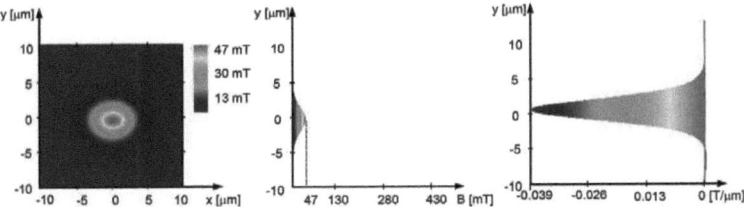

Figure 6.9: Magnetic field and gradient at 2500 nm. At distances equal to the radius of the tip size the field and the filed gradient change with magnetic dipole behavior. At 2.5 μm the maximal field generate is 47 mT and the field gradient is 0.039 T/ μm.

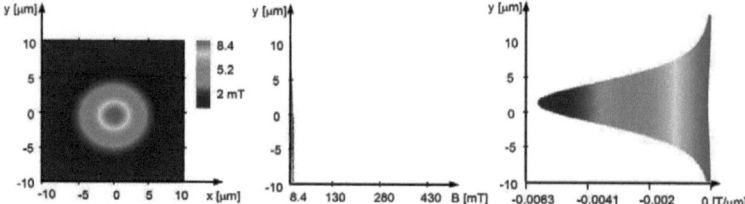

Figure 6.10: Magnetic field and gradient at 5500 nm. At distances of 5.5 μm the field and the field gradient have magnetic dipole behaviour. At 5.5 μm the maximal field is 8.7 mT and the field gradient is 0.0063 T/ μm.

At a distance larger than the size of the magnetic particle, the magnetic field generated by the tip can be assumed to be a magnetic dipole. At a short distance the field must be calculated with the real tip size for the strong inhomogeneity caused by the edges.

The dipole model strongly simplifies the calculation of the magnetic field, of the magnetic gradient and of the tip sample interaction. The magnetic field and the field gradient produced along the z axis by a magnetic dipole with radius 2.02 μm and by the magnetic particle are represented in figure 6.11 as a function of the distance. At long distance the spherical tip shows the best fit with the tip particles mounted on the mechanical lever.

The difference of the magnetic field amplitude and magnetic field gradient between spherical tip and magnetic particle become important for distances smaller than the magnetic particle. This effect is caused by the edges that cause a fast decay of the field. For distances larger than the tip size the magnetic field can be approximated with the magnetic dipole behavior. This approximation for its spherical symmetry has an analytical solution. The equation of the magnetic field and the gradient filed in z direction[1] is:

$$B_z = \frac{1}{3}\mu_0 \frac{a^3 M(r^2 - 2z^2)}{(r^2 - z^2)^{\frac{5}{2}}} \tag{6.13}$$

and

$$\frac{dB_z}{dz} = \frac{1}{3}\mu_0 M a^3 \frac{r^2 z - 6z^3}{(r^2 - z^2)^{\frac{7}{2}}} \tag{6.14}$$

and the forces generate are:

$$F_z = \iiint_{V_s} \frac{\partial M}{\partial z} B_z dV = \int_0^\infty \int_0^{2\pi} \frac{\chi_0 B_z^2}{2\mu_0} r d\phi dr = \frac{1}{24} \frac{\chi_0 a^6 B_z^2 \pi}{\mu_0 d^4} \tag{6.15}$$

---
[1] Cylindrical coordinate.

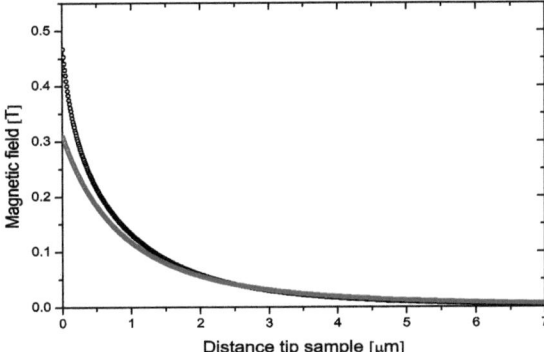

Figure 6.11: Magnetic dipole field. The magnetic field induced by the magnetic tip can be modeled with a magnetic dipole. At smaller dimensions the edge effects modify the behavior of the magnetic field. The graph represents the induction of the magnetic field in function with the distance. The black line represents the magnetic field generates by the magnetic particle. The red line the best magnetic dipole fit (radius 2.02 m).

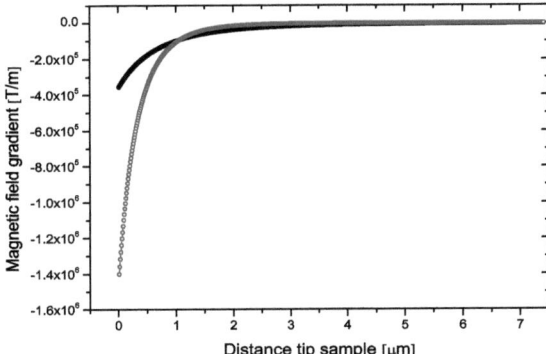

Figure 6.12: Magnetic dipole gradient. The graph represents the gradient in the z direction. The black line represents the gradient generates by the particle and the red line one represents the gradient generate by the magnetic dipole.

Figure 6.11 represents the diamagnetic force calculated with equation 6.10 generated by the magnetic dipole sphere approximation on quartz and on HOPG infinite sample. Due of the susceptibility of the samples different repulsive interactions forces are induced.

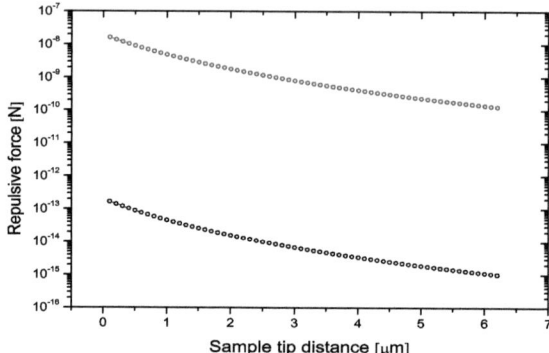

Figure 6.13: Diamagnetic force as a function of the distance. The graph shows the diamagnetic force as a function of the distance for a magnetic sphere with radius 2.02 μm with permanent magnetization of 0.986 T and a sample of quartz (bottom line). The force calculated is in the 10^{-14} N range. The force generated by a magnetic sphere with radius 5 μm and a HOPG sample is 10^{-9} N.

In the long distance range, the diamagnetic interaction force between the vertical cantilever and the quartz sample has amplitude of 10^{-14} N. This tensile force causes a frequency shift on the order of a mHz, so 1000 times smaller than the frequency measured in figure 6.12. The effect becomes important only for forces equal or larger than 10^{-9} N. The vertical cantilever would be sensitive to this diamagnetic force only when the contact potential is compensated and when no lateral gradient are present.

6.2 Measurement: tip-sample interactions

In this section we extrapolate the forces and force gradients from the frequency shift experimentally measured. In the previous section we have theoretically calculated the influence of the different forces, which induces a frequency shift. The total frequency shift measured is the sum of the different interactions forces effects. The frequency shift of the horizontally cantilever, which is not sensitive of the static vertical force, increases due to a repulsion force gradient and decreases due an attractive force gradient. The force gradient interaction can be consequently extrapolated directly from the frequency shift. On the other hand the frequency shift of a vertical cantilever depends of the sum of the lateral force gradient interactions and the vertical force. Consequently, the frequency shift change has to be analyzed in order to determine the effect or the effects that simultaneously interact.

As in a standard MFM, a cantilever tipped with a modeled 5 μm magnetic sphere horizontally approaches a HOPG sample. A strong repulsion force gradient is measured at a 3 μm distance over the sample. This force gradient causes a frequency shift of 10 Hz. The graph shows that the electrostatic forces is purely attractive, in fact by applying a potential the cantilever is attracted from the sample and the diamagnetic repulsive force is canceled. A pure diamagnetic force is measured when the contact potential is compensated (contact potential -1 V).

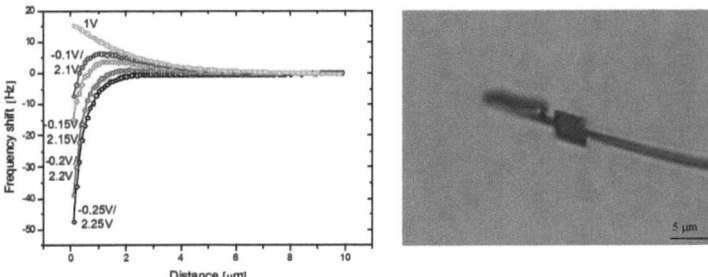

Figure 6.14: Frequency shift as a function of the distance in horizontal approach.
The graph represents the horizontal approached IBM cantilever as a function of the distance with different potential applied (the sample is grounded). When the contact potential is compensated the frequency shift can be fitted with the diamagnetic forces gradient. The sample is HOPG with a susceptibility $\chi_\perp = 450 \times 10^{-6}$.

A similar graph is measured when a tipped perpendicularly cantilever approaches a quartz sample. The positive frequency shift is induced in this case not from a diamagnetic repulsive force, but from an attractive electrostatic force. The diamagnetic force in fact induces frequency shifts in the mHz range, as explicate in section 6.1.4. The frequency shift increases due to the strong electrostatic force, as represented in the following graph. The more the cantilever approaches the sample the more important the horizontal forces gradient become due of the edges effects. So the frequency shift decreases due to the tilted alignment or sample inhomogeneity.

In order to understand the causes of the frequency shift induced by the lateral gradient, we should scan the sample at a constant height. Dipole molecules could in fact induce either positive or negative force gradients, which increase or decrease the frequency shift. Therefore it would be interesting to repeat the experiment with an annealed sample, where the surface it perfectly clean.

The frequency shift that is measured with a vertical cantilever and represented in graph 6.15 can be explained with two electrostatic interactions. At long distances the attractive tensile force increases the frequency shift. At small distances the lateral force gradient due of the magnetic particle geometry and the charge distribution reduces the frequency shift. These two effects interact and cause not only a change of the frequency shift but also a change of the Q factor, which is represented in figure 6.15.

At distance larger than the particle size, the Q factor decreases linearly with the distance. At distances smaller than the particle size, the edges effect and the inhomogeneity become important and the behavior of the Q factor is slightly changed. We observe in fact an increase of the Q factor as a function of the distance.

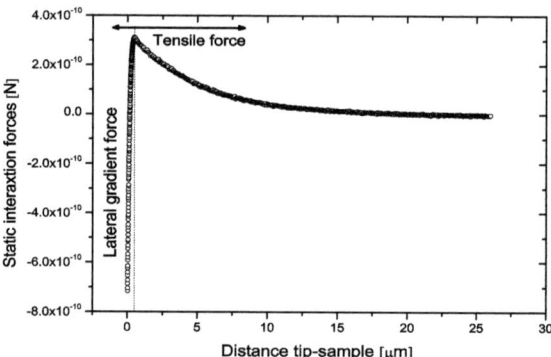

Figure 6.15: Frequency as a function of the distance. The graph represents the attractive force and the lateral force gradient acting on the cantilever.

In the long-range distance, the dissipation process increases as the cantilever approaches the sample, because the dissipation is caused by the electric field fluctuation. Assuming that the charge fluctuation causes the dissipation, we can calculate the total moving charges. The charge fluctuation friction is calculated subtracts the total friction with the internal elastic friction with the following relation:

$$\Gamma_s = \Gamma - \Gamma_0 = \frac{k_0}{\omega_0^2}\frac{\omega}{Q} - \frac{k_0}{\omega_0}\frac{1}{Q_0} \qquad (6.16)$$

where the Q_0 factor is measured at an infinite distance and the Q factor is measured while approaching the cantilever to the sample. The measured friction as a function of the distance is reported in figure 6.17. The graph proves an increase of the friction for distances larger than the tip size. The decrease of the friction at a few microns is consequently due of the geometry of the magnetic particle and the inhomogeneity of the charge distribution.

As the cantilever approaches to the sample surface, the capacitance between magnetic particle and sample increases. This increases the so called "KTC" noise, which refers to the total charge fluctuation stored in the capacitance. The friction is then increased, because it is quadratically dependant on the capacitance. This behavior is respected at distances larger than the particle size. At distances smaller than the particle size edge effects must to be considered.

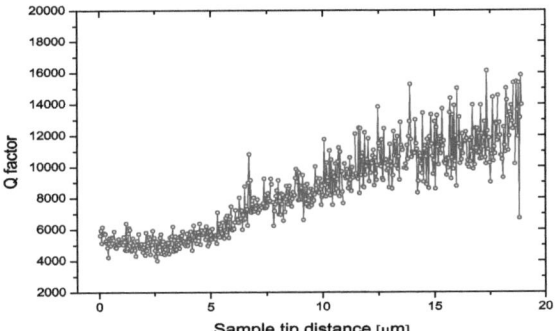

Figure 6.16: Q factor as a function of the distance. The graphs represent the attractive force and the lateral force gradient acting on the cantilever.

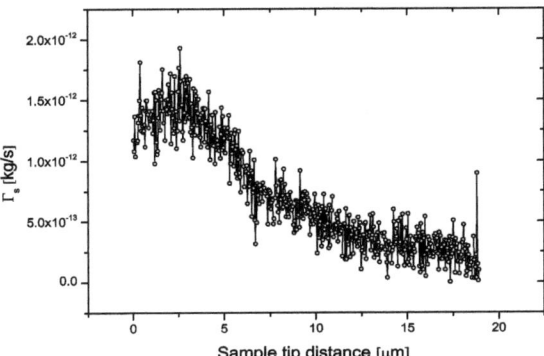

Figure 6.17: Friction as a function of the distance. The graph represents the friction due to the charge fluctuation between two plates of the capacitor (tip and sample) as a function of the distance. The oscillation of the cantilever causes a movement of the tip plate causing a field variation when the potential is constant. The variation of field causes a variation of the charge.

Comparing the power spectral density of the force and the power spectral density of the force fluctuation we find the relation 6.14.

$$S_E = \frac{\Gamma_s 4 k_B T}{C^2 V^2} \qquad (6.17)$$

The formula relates the power spectral density of the electric field and the friction calculated with relation 6.13. When the potential and the capacitance between sample and tip are known, we can calculate the fluctuation charges, that causes the frictional loss. The fluctuation charge is calculated with the power spectral density of the electric field and the derivative of the capacitance between sample and tip. The relation of the fluctuation charge is

$$\delta_e = \frac{E}{e} \frac{\partial \frac{1}{C}}{\partial x} = \frac{\sqrt{S_E} 2\pi \varepsilon_0 R}{e \left(\frac{1}{\ln d} - \frac{R+d}{\ln(d^2)} \right)} \qquad (6.18)$$

where e is the elementary charge, R is the radius of the magnetic tip, d is the distance tip-sample and E the electric field. The capacitance is calculated with the relation 6.3 , which assumes an infinite plane and a spherical tip through a constant potential. The figure 6.18 represents the square of the power spectral density of the electric field and the charge spectral noise fluctuation as a function of the distance.

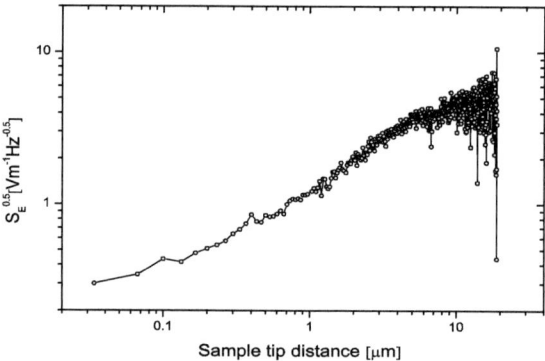

Figure 6.18: Power spectral density as a function of the distance. The graph represents the electric field spectral density as a function of the distance. At small distances the electric field fluctuation become smaller.

As showed in figure 6.18, the square of the power spectral density of the electric field has a linear dependence on the distance. The electric field has a linear dependence on the capacitance. At large distances, the electric field is not defined as precisely as at micron distances, because the induced friction is very small. The charge spectral density has a quadratic behavior as a function of the distance, because the charges are correlated with the power of the capacitance. Due of the

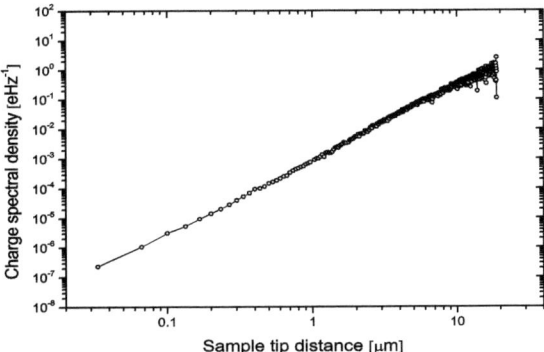

Figure 6.19: Charge spectral density as a function of the distance. The value reported at nanometer range has a charge spectral density similar to the one reported by the single electron transistor experiment.

largeness of the magnetic particle, the calculated charges noise cause an important friction processes. This can reduce the sensitivity of the cantilever at smaller distances.

The inhomogeneity charge distribution over the sample and the cantilever, the geometry of the magnetic particle and the dipole molecules attached to the sample surface can cause a change in the slope. An important change in the magnetic dissipation is detected at particle size distance, where the real geometry and charge distribution should be used.

Stipe and coworkers already have measured the charge spectral density caused by the electric field fluctuation [11]. At 125 nm tip to sample distance they found a charge spectral density of 2×10^{-5} e $Hz^{-0.5}$ a value that they indicate on the level of the observed charge fluctuation in the single-electron transistor. Our data shows a comparable value. We represent in figure 6.19, the spectral charge noise as a function of the distance. When the constant potential is compensated the dissipation is reduced, because the charge induced by the capacitance is reduced.

Chapter 7

Signal to noise ratio in continuous wave magnetic resonance force microscopy

Nuclear Magnetic Resonance (NMR) and Electron Spin Resonance (ESR) [31], [32], [33] are the resonance effects, which allow us to reconstruct the molecular structure by resonance absorption. The correspondence between NMR and ESR is very close, and much of the basic theory of NMR is directly applicable to ESR. In both ESR and NMR it is also necessary to provide an external static magnetic field to establish the ground state and excited state energy levels. In the case of ESR it is necessary to have an unpaired electron instead of an unpaired nuclear spin as in NMR. The major difference between the two techniques is due to the gyro magnetic ratio for the proton and electron. ESR spectroscopy has a higher absorption frequency than NMR spectroscopy. Consequently, the sensitivity of EPR is considerably higher approx. by a factor of 1000. In addition the absorptions lines for ESR are also significantly broader.

Recent efforts to develop microNMR and microMRI have been based on pulsed techniques. In the limit of extremely small samples structures have employed the mechanical resonance of cantilever such as are those used in atomic force microscopy (AFM). Research groups have tried to increase the signal to noise ratio (SNR) in standard NMR and ESR by increasing the magnetic field and consequently by optimizing the microcoil antenna [34],[35],[36]. An ESR sensitivity of 10^7 spins has been achieved. Therefore it seems reasonable to reexamine the fundamental detection limits for ESR in view of advances in AFM technology. Rugar and coworkers have demonstrated a sensitivity of 1 spin [26].

In this chapter, we present a comparison of inductive continuous wave (CW) and mechanical magnetic resonance detection for the case of micro ESR. New techniques of pulse excitations have been proposed and the reality of a new complete scanner for nm range seems to be feasible.

7.1 Theory: Signal to Noise Ratio in electron spin resonance

Since the first experiments of Bloch [37], [38], [39] and Purcell [40] in 1946, the detection of nuclear magnetic resonance in bulk matter has usually been carried out by measuring of the electromotive force s(t) induced by the nuclear magnetization precession in a coil near a sample under investigation. In addition to the electromotive force, a random noise signal n(t) will also be present. The total voltage across the coil x(t) can be expressed by as

$$x(t) = s(t) + n(t) \tag{7.1}$$

The electromotive force ds(t, $\mathbf{r}_s$) induced in a coil placed in proximity to the volume sample dV_s in an applied magnetic field $\mathbf{B}$(t, $\mathbf{r}_s$) can be deduced from the vector dot product of $\mathbf{B}$(t, $\mathbf{r}_s$) and $\mathbf{M}$(t, $\mathbf{r}_s$) by the principle of reciprocity [34] as

$$\delta s(t, \vec{r_s}) = -(-\frac{\delta}{\delta t}(\vec{B_{ud}}(\vec{r_s}) \cdot \vec{M}(t, \vec{r_s}))dV_s) \tag{7.2}$$

where $\mathbf{B}_{ud}(\mathbf{r}_s)$ is the magnetic field at position $\mathbf{r}_s$ produced by a unit direct current (DC) carried by the detection coil. This relation, derived directly from the Faraday and Biot-Savart laws, holds exactly for a non-conducting sample placed within or at a small distance from the detection coil.

For millimeter diameter or smaller coils, the thermal noise associated with the resistance of the detection coil is much greater than dominates the noise from the sample [35]. The root mean square value (rms) of the thermal noise n_{rms} associated with a resistance R_c is given by

$$n_{rms} = \sqrt{4k_b T_c R_c \triangle f} \tag{7.3}$$

where T_c is the temperature of the coil, k_b is Boltzman's constant and Δf is the bandwidth of the amplifier.

The signal-to-noise ratio (SNR) can be defined as the square of the ratio between the power spectral density of the signal detected to the power spectral density of the thermal noise. By using a matched filter for the detector [41] the SNR ratio is maximized and becomes

$$SNR = \sqrt{\frac{\int\limits_{-\infty}^{\infty} s(t)^2 dt}{2\pi \left(4k_b T_c R_c\right)}} \tag{7.4}$$

where the numerator S(ω) is the Fourier transform of the signal s(t) and the denominator W_N (ω) is the power spectral density of the noise n(t).

Therefore, the SNR of a conventional ESR system can be calculated if we know the power distribution of the absorption signal energy. The signal absorption line of a resonance effect takes the shape of a Lorentzian function given by $1/\pi(T_1/(1+A\omega^2 T_1^2))$ in term of the angular frequency ω and the spin lattice relaxation time T_1 (the longitudinal relaxation time). The absorption line width Δv is given by $\Delta v = 1/o\pi T_2$, which depends on the spin-spin relaxation time T_2 (the transverse relaxation time). The minimum noise ground floor reachable corresponds to the Brownian noise and is caused mainly by the coil resistor matched at 50 Ω.

For the slow passage experiment the induced magnetization **m**(t,**r**) can be calculated using the steady-state solution (saturation) of the empirical vector Bloch equation [38], [39], [42] where $T_2 \ll$ sweep rate and $T_2 \approx T_1$.

$$\frac{\partial \vec{M}}{\partial t} = \gamma \vec{M} \times \vec{B}_{ext} - \frac{M_{rx}\vec{i} + M_{ry}\vec{j}}{T_2} - \frac{M_z - M_0}{T_1}\vec{k} \quad (7.5)$$

where **M** is the vector magnetization of the sample, and M_{rx}, M_{ry}, M_z are the components of **m** in the rotating frame **i**,**j**,**k**, M_0 equilibrium value of the magnetization, T_1 the longitudinal relaxation time ("spin-lattice relaxation time"), T_2 the transverse relaxation time ("spin-spin relaxation time") and $\mathbf{B}_{ext}$ the total external magnetic field. In the steady state condition the change of the magnetization components lead to:

$$\frac{\partial M_{rx}}{\partial t} = \frac{\partial M_{ry}}{\partial t} = \frac{\partial M_z}{\partial t} = 0 \quad (7.6)$$

Consequently, if a long time has elapsed for the transient exponentials to decay, we can write:

$$M_{rx} = \frac{(\omega - \omega_0)\gamma B_1 T_2^2}{1 + (T_2(\omega - \omega_0))^2 + \gamma^2 B_1^2 T_1 T_2} M_0 \quad (7.7)$$

$$M_{ry} = \frac{\gamma B_1 T_2}{1 + (T_2(\omega - \omega_0))^2 + \gamma^2 B_1^2 T_1 T_2} M_0 \quad (7.8)$$

$$M_z = \frac{1 + ((\omega - \omega_0)T_2)^2}{1 + (T_2(\omega - \omega_0))^2 + \gamma^2 B_1^2 T_1 T_2} M_0 \quad (7.9)$$

We remark that the components M_{rx} and M_{ry} rotate at the Larmor frequency in the perpendicular plane around the direction of the static magnetic field. These two components are used in the inductive mode in order to induce the variation of electrical potential in the RF coil, which is placed in the x,y plan. Hence it is unable to detect a signal corresponding directly to change in M_z without any particular pulse technique. On the other hand, the MRFM detection is unable to detect the x,y components, nerveless the cantilever vibrates at the Larmor frequency. Consequently, the z static component is always used for couple the signal to the mechanically sensitive beam.

7.1.1 Inductive coupled continuous wave Electron Spin Resonance Signal to Noise Ratio

In this subsection we extrapolate the signal to noise ratio of a standard inductive continuous wave electron spin resonance spectrometer. The coil is placed in the x,y plane in order to detect the components M_{rx} and M_{ry} in the assumption of slow passage.

In a typical continuously wave ESR experiment, the shape of the magnetic field variation is usually a sinusoidal signal. This kind of signal induces a deformation of the shape of the absorption line due of the non linear behavior. Consequently, a saw-tooth function signal is used for varying the magnetic field as shown in figure 7.1. The graph sketch four sweeps with a large magnetic field scan. In a optimal setup the scan largeness must be on the order of the absorption line width.

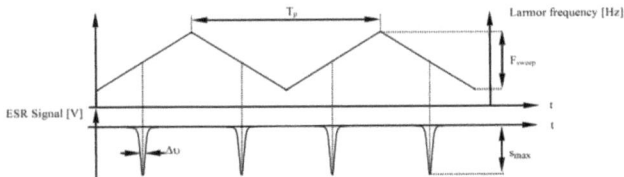

Figure 7.1: Continuous wave electron spin resonance experiment. The graph on the top shows the magnetic field sweep function, e.g., a saw-tooth, while the graph below corresponds to the absorption ESR signal. The total time of the detection is given by T_t.

The amplitude and line shape of the ESR signal are strongly dependent upon the sweep range F_{sweep}, the relaxation times T_1 and T_2, the time period T_p, the sweep rate $R_{sweep}=2F_{sweep}/T_p$ [Hz/s] and the RF magnetic field strength B_1. At high sweep rates and in strong static magnetic fields the ESR signal becomes broader and wiggles appear in the base line [43]. The ESR signal s(t), based on the analysis of Ernst and Anderson, can be approximate to the relation 7.10, even when it become a strongly distorted Lorentzian function [44].

$$\int_{-\frac{T}{2}}^{\frac{T}{2}} s(t)^2 dt \cong \frac{\pi}{4} \frac{\Delta v}{R_{sweep}} s_{max}^2 \tag{7.10}$$

If time averaging over n acquisition is assumed and since $n=2T_t/T_p$ where T_t is the total time for signal detection and $T_p=2F_{sweep}/R_{sweep}$, the SNR per unit time is given by

$$\frac{SNR}{\sqrt{T_t}} = \sqrt{\frac{\Delta v}{32 k_b T_c R_c R_{sweep}}} s_{max} \tag{7.11}$$

where s_{max} and Δv are the maximum and the full width at half maximum of the ESR signal, respectively. In the case of slow passage, The SNR can be estimated with a field sweep through resonance. In fact, for a slow passage (i.e. for $(\gamma B_x)^2 T_1 T_2 \ll 1$), the maximum signal is given by

$$s_{max} = \omega_0 M_0 B_u V_s \frac{\gamma B_1 T_2}{1+\gamma^2 B_1^2 T_1 T_2} \tag{7.12}$$

where ω_0 is the Larmor frequency and M_0 is given by a general expression of the Curie law [42],

$$M_0 = N_{atoms/volume} \frac{\gamma^2 \left(\frac{h}{2\pi}\right)^2 B_0 I_s (I_s+1)}{3 k_b T_c} \tag{7.13}$$

where h is the Planck constant, γ the gyro magnetic ratio, k_b the Boltzman constant, I_s the electron spin number which is equal at $1/2$ for an electron and T_c the temperature.

Finally the maximum SNR ratio is given by

$$\frac{SNR}{\sqrt{T_t}} = 0.12\sqrt{\frac{1}{4k_bT_cR_c\Delta f}}\omega_0 M_0 B_u V_s \quad (7.14)$$

If we assume that the sample is enclosed within a single-layer, of a very long solenoid [45], [46], then the RF field at the center is homogenous and can be defined by B_u per unit current, also called the sensitivity of the coil.

The experimental results using microcoils report that to optimize the detection sensitivity the microcoil diameter d_{coil} must be reduced, the sample scaled down and the static magnetic field increased as the frequency [35]. At small coil dimensions and at high frequency the proximity effect and the skin depth begin to play an important role. The ESR sensitivity per unit volume has consequently a different relation with the coil dimensions (value reported in the table 8.1).

The RF coil is frequency dependent and the limit of frequency range is given by the RLC circuit, where L is the inductance of the coil, R the parasite resistance and C the parasite capacitance of the coil. Frequency of 1.5 Ghz is the working limit of the solenoidal coil. Higher frequency could be reached using a strip line antenna, where the eigenfrequency is pushed to ward the GHz range.

It is well know and accepted [36] that aspect ratio and the filling factor can increase the SNR ratio per unit volume by a factor of 10 between a microcoil with a diameter of 1 mm and one of 10 μm. In our experiments, we have taken the dimensions of the coil and the amplitude of the static magnetic field as fixed and we have varied the sample quantity until the systems limits of the detection reached.

Static magnetic field	Temperature	Gyromagnetic ratio of atom or electron	Time of the measurement	Dimension coil fixed (only the sample is reduced)	Maximum aspect ratio and filling factor
B_0^2	$1/T^{\frac{3}{2}}$	γ^3	$\sqrt{T_t}$	d_{coil}^3	d_{coil}^2 (d_{coil} for >200µm) 5 d_{coil}^2 (d_{coil} for <200µm)

Table 7.1: Inductive sensitivity. The table shows the SNR of a standard NMR as a function of temperature, sample dimension, coil dimension and finally average time.

The table 7.1 is extrapolated from the equation 7.14. The varying parameters are assumed to be independent and under the assumptions of slow passage. The SNR in inductive detection is strongly dependent of the static polarizing magnetic field and the coil geometry.

7.2 Magnetic resonance force microscopy and signal to noise ratio

In a typical MRFM continuous wave experiment, where the sample is placed on the cantilever, the SNR is calculated using an equation corresponding to equation of 7.14.

The main differences between MRFM and ESR inductive techniques are due to the z sensitive component. The cantilever is sensitive to the z force gradient s(t) signal with a power spectral density $S(\omega)$ and to the Brownian thermal motion noise n(t) with a power spectral density $W_N(\omega)$. Moreover, the MRFM needs a magnetic field gradient that interacts with the polarized sample and generates a constant signal force producing a deflection of the beam cantilever. The z force is switched on and off, by the RF coil excitation and by the variable magnetic polarizing field. The signal in MRFM is proportional with the force detected by the cantilever [19] and this is expressed by:

$$\vec{F} = \nabla(\vec{m} \cdot \vec{B}) \qquad (7.15)$$

where $\mathbf{m} = \mathbf{M}_s \mathbf{V}_s$ is the magnetic moment due to electron spins, $\mathbf{M}_s$ is the magnetization, $\mathbf{V}_s$ is the volume of the sample and $\mathbf{B}$ is the magnetic field over the volume sample. The force over the sample in a magnetic linear gradient can be calculated by integrating the differential force over the sample as given by:

$$F_z = \int_{sample} M_z(B_z, B_1) \frac{\partial B_z}{\partial z} dV \qquad (7.16)$$

If we assume that the magnetization is given by solution to the steady-state Blockh equation, then M_z can be expressed by the equation 7.9 with the gradient and static magnetic field.

The maximum change of magnetization is given by the difference of the entire sample when it is in resonance and out of resonance. The change of magnetization is,

$$M_z = M_z^{max} - M_z^{min} = \frac{M_0 \gamma^2 B_1^2 T_1 T_2}{1 + \gamma^2 B_1^2 T_1 T_2} \qquad (7.17)$$

The absorption line width of the solid diphenylpicrylhydrazil (DPPH) sample is extremely wide. It is 0.5 mT at low magnetic field [47], [5]. Consequently, the entire sample can be excited with a gradient magnetic field having a maximal change of the magnetic field of 0.5 mT through the entire sample. This should induce the strongest force change expressed by:

$$F_z = \frac{M_0 \gamma^2 B_1^2 T_1 T_2}{1 + \gamma^2 B_1^2 T_1 T_2} \frac{\partial B_z}{\partial z} V_s \qquad (7.18)$$

As in the case of the thermal resistor, the cantilever noise is due to the Brownian motion [48], [49] of the oscillator. Thermodynamic analysis shows that the rms fluctuation of a force-noise oscillator is the average given by the relation [48]:

$$F_{n_{rms}} = \sqrt{4k_b T_c \alpha \Delta v} = \frac{\sqrt{0.32 \cdot 4k_b T_c k}}{\omega_m \cdot \tau} \qquad (7.19)$$

The bandwidth $\Delta v = 0.32/\tau$ is essentially the bandwidth of the mechanical oscillator and is given by the Lorentzian shape of the mechanical oscillator. The damping constant $\alpha = m_{eff}\omega/Q = k/(\omega_m^2 \tau)$ is defined by the effective mass m_{eff} by the Q factor and by the unloaded resonance frequency oscillation. The ω_m defines the loaded cantilever mechanical resonance frequency and, τ the damping time. Therefore, in summary the SNR for this detection can be expressed by:

$$\frac{SNR}{\sqrt{T_t}} = \frac{\omega_m \cdot \tau}{\sqrt{0.32 \cdot 4k_b T_c k}} \cdot M_0 \cdot \frac{\gamma^2 B_1^2 T_1 T_2}{1 + \gamma^2 B_1^2 T_1 T_2} \cdot \frac{\partial B_z}{\partial z} \cdot V_s \qquad (7.20)$$

The table 7.2 is extrapolated from the relation 7.20 and resumes the SNR behavior as a function of the different factors. The entire sample is assumed to be in resonance and in a constant gradient field. The SNR is linearly dependent on the magnetic field and with the mechanical geometry of the cantilever.

Static magnetic field	Temperature	Gyromagnetic ratio of atom or electron	Time of the meas.	Damping time	Mechanical resonance frequency	Spring constant
B_0	$1/\sqrt{T_c^3}$	γ^2	$\sqrt{T_t}$	τ	f_m	$1/\sqrt{k}$

Table 7.2: Force sensitivity. The table shows the SNR behavior of the MRFM as a function of the temperature, average time, mechanical resonance frequency and spring constant.

Comparing the table 7.1 and 7.2, it appears that the SNR behaves differently as a function of the magnetic field and geometric factors. The force sensitivity sensor is only linearly proportional to the static magnetic field instead of being quadratically proportional as is the case with the inductive methods. Furthermore, the measured components are different. The MRFM sensor is sensitive to the m_z component, while the inductive method is sensitive to the m_x and the m_y components.

Chapter 8

Electron spin resonance at room temperature

In this chapter, the theoretical results discussed in the chapter 7 are used in order to measure the ESR signal with the inductive methods and with the force magnetic resonance force methods at room temperature. Different quantities of samples volume are introduced in the coil and are mounted directly on the cantilever at a constant magnetic field. The SNR is then measured and compared. The results shows that the MRFM machine can considerably improve the sensitivity of ESR and NMR sample at room temperature.

In chapters 4,5 and 6 the cantilever mount the permanent magnetic particle, which provides the gradient that induces the resonance slice in the nearby sample. This setup has a high sensitivity only at magnetic particle distances, due to the strong gradient generated by the permanent magnet. The experiment that we propose in this chapter want to excite an entire micrometer sample and not merely part of it. For this reason, we invert the position of the sample and of the permanent magnet. In fact the sample is mounted directly on the cantilever and on the permanent magnet, which provides the gradient, approached at millimeter distances. This setup makes possible to excite the entire sample mounted on the lever by adjusting the distance of the permanent magnet.

8.1 CW-ESR Inductive Experiments

In order to verify the SNR model discussed in chapter 7, we conducted a series of experiments using a paramagnetic material with a very short T_1, which maximizes the detected signal. We selected the paramagnetic crystal of diphenylpicrylhydrazil (DPPH[1]) as a sample for its very short T_1, approximately 60 ns ($T_2 \cong T_1$) [50].

In this study six capillaries with diameters from 100 μm to 1 mm were filled with DPPH and introduced in a 8 turns solenoidal coil with a length of 2.2 mm and inner an diameter of 1 mm. The coil is then tuned at 1.4 GHz for a static magnetic

---

[1]DPPH, 2.2-Diphenyl-1-picrylhydrazyl, D9132, Fluka Chemie Sarl, Industriestrasse 25, 9470 Buchs, Switzerland. http://www.fluka.ch

field of 50 mT. A scheme of the ESR excitation and detection system is shown in figure 8.1.

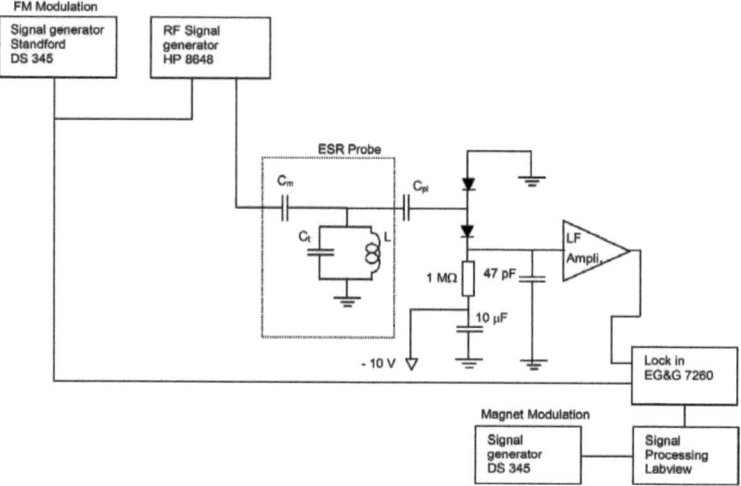

Figure 8.1: Inductive ESR scheme. The scheme diagram for the circuit used to detect the CW-ESR signal. The RF modulated signal frequency is decoupled from the detector amplifiers using 2 Schottky diodes. When an ESR signal is detected, the signal is low pass filtered and amplified with a factor of 25000 before demodulated with the lock-in amplifier.

The signal generator[2] was used to generate the RF signal at the Larmor frequency of 1472.15277 MHz. The RF signal is modulated[3] at a frequency between 200 Hz and 30 kHz. Another signal generator is used to sweep in a saw tooth manner the magnetic field with a frequency of 2 Hz. In order to isolate the RF amplifier from the excitation signal a pair of a Schottky diodes[4] were introduced. The acquired RF signal was amplified with a gain of 25000 and a bandwidth of 3 Hz to 30 kHz. Then the signal is amplified using a lock-in amplifier[5] and collected using a Labview A/D conversion card[6]. The collection data is done by measuring the real (x) and imaginary (y) part of the lock-in amplifier.

The measurements were performed using six DPPH samples that ranged in a volume from 0.0078 to 0.8 mm^3 (table 8.1). Since DPPH has a molar mass of 394.3 g/mol, a spin density[7] of 2.3 x 10^{27} spin/m^3 [51], [52] and a mass density of 1506

---

[2]Hewlett-Packard, 8648
[3]Stanford DS345
[4]Siemens, Bat-64
[5]E&G, 7260
[6]National Instruments, PCI-mio-16e-1
[7]In the thesis we will use in order to compare the continuous wave spectrometers with the magnetic resonance force microscopy a spin density of 6.57 x 10^{24} spin per m^3.

kg/m^3, the samples should contain between 1.8 x 10^{18} to 17 x 10^{15} spin/m^3.

The signal measured do not confirm this density. In fact, if we calculate the effective signal produce from these samples with the thermal noise we find a ratio between the theoretical sample and real sample of 350. Consequently, the spin density is 6.57 x 10^{24} spins per m^3. This difference is probably due to the oxidation of the DPPH and of its crystal structure. The samples contain between 5.1 x 10^{15} to 5 x 10^{13} spins.

ESR sample	Sample1	Sample2	Sample3	Sample4	Sample5	Sample6
φ[mm]	1	0.8	0.6	0.4	0.2	0.1
L[mm]	1	1	1	1	1	1
V[mm^3]	0.79	0.5	0.28	0.13	0.03	0.0078
Spins	5.1x10^{15}	3.1x10^{15}	1.9x10^{15}	8x10^{14}	2x10^{14}	5x10^{13}

Table 8.1: Sample measured by inductive methods. Six samples of DPPH have been introduced on six capillaries and the quantity of spins calculated.

The absorption line signal generated by the DPPH sample 5 is represented in figure 8.2. The magnet is sweep at 20 Hz with an amplitude of 3-6 mT and a static magnetic field of 50 mT.

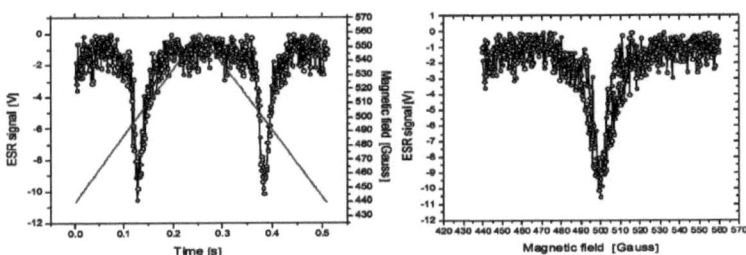

Figure 8.2: Absoption line of DPPH. The DPPH signal, obtained using the sample 5, shows a signal to noise ratio of about 10 for a single sweep, one with increasing field and one with decreasing the field. The magnet is sweep at a frequency of 2 Hz in a range of 100-200 Gauss. The line width of absorption of the DPPH is between 5 to 10 Gauss.

The signal represented in figure 8.2 is the absorption line during a sweep. The noise floor is 10 times higher than in the optimal theoretical case. The thermal noise measured is 0.4 V that corresponds for an amplification of 25000 to 1.6 μV. This is around 10 times the theoretical thermal noise of 0.14 μV that we should measure after the diode for a 50 Ω matched-coil. The additional noise is due to intrinsic noise of the diode and of the external power supply. The noise is reduced by decoupling the DC amplifier from the electrical nets. Moreover, the diode could be removed by introducing two orthogonal coils one used for exciting the sample and one for decoupling the detection signal from the excitation.

The SNR of the home built system does not have the sensitivity reached with commercial system at higher polarizing field. The best sensitivity reached known is of 10^{-7} spins [3]. The room temperature machine shows at standard condition and at low magnetic field a sensitivity of 10^{-11} spins. This sensitivity is easily reached and improved with a standard MFM machine.

8.2 CW magnetic resonance force microscopy

The measurements that are performed in the previous subsection are repeated with the home built MFM microscope. The measurements show the improvement of sensitivity and consequently the potential of the MFM machine in ESR and NMR spectroscopy.

In figure 8.3 is represented the room temperature head of the microscope used for the CW-ESR experiment. Nanogram sample of DPPH are attached to a micro cantilever positioned near a permanent magnet (Sm_2Co_{17}) that induces a strong magnetic field gradient (5 T/m) at a distance of 100 micrometers. An external electromagnet is swept very slowly (0.001 Hz). The RF coil, which generates an oscillating magnetic field at 1.4 GHz, is frequency modulated at the first eigenmode of the mechanical lever. At resonance condition, the radio frequency turns off the attractive static force modulated with the frequency of the mechanical lever. An interaction is then measured by a lock-in amplifier.

Figure 8.3: Photograph of the MRFM microscope's head. On the left it is possible to see the tuning capacitor and the RF coil (tuned to 1.4 GHz), in the center is the permanent magnet used to establish the gradient in B_0 (Sm_2Co_{17}), the cantilever, and the piezo actuator.

The electron spins are excited and flipped to the xy plane, when the radiated RF frequency and the magnetic field correspond to the gyromagnetic ratio of electron 28 GHz/T. In the resonance condition, the spins are positioned perpendicular to the magnetic field and no force is acting on the cantilever. Off resonance, the force acts on the cantilever, but it is not periodically turned off by the RF field. The magnetic resonance signal is acquired using a lock-in amplifier and digitized by a Labview module as show in the block diagram of the electronics used to perform the experiment (figure 8.4).

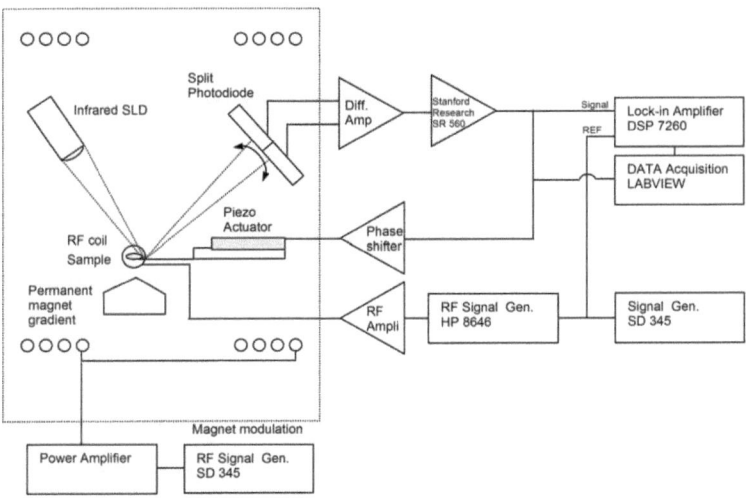

Figure 8.4: Scheme of the MRFM detection. The magnetic field B_0 is established and modulated by an electromagnet. The DPPH sample is attached on a commercial cantilever. The cantilever is placed in the center of the electromagnet near a gradient arising from to a permanent magnet. A coil generates an RF field at 1.4 GHz that is modulated at the mechanical frequency of the cantilever. When the spins of the sample are rotated in the plane of the RF coil no force acts on the cantilever. It is possible to detect the change in the force acting on the cantilever by using a 4-quadrant detector.

In other to determine the mass attached on the cantilevers and consequently the number of spins excited, the mechanical resonance frequency f_0, the Q factor of all Nanosensors cantilevers were measured before attaching the sample. The resonance frequency is measured using a Fourier Fast Transformation (FFT) Labview module program able to excite the cantilever and measure the frequency at a precision of less than 0.1 Hz. The mass is then deduced by the frequency shift induced by the mass attached. The commercial cantilevers measured have on average a mechanical resonance frequency of 10.7 kHz and a spring constant of 0.16 N/m.

In order to perform the ESR-MRFM experiment, the DPPH samples were physically attached using optical glue[8] to seven different Nanosensors cantilevers. The ESR MRFM apparatus is sketched in the block diagram in figure 8.4.

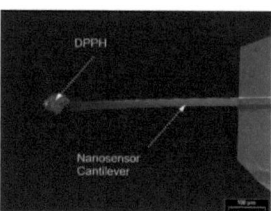

Figure 8.5: Photograph of sample. SEM picture of 10 kHz Nanosensor cantilever with sample 3 (530 ng) of DPPH mounted on the end.

After attaching the samples, the resonance parameters of the sevens cantilevers were again measured and the mass of the sample determined. All cantilevers were photographed (see example figure 8.5) with the Scanning Electron Microscope (SEM) and the volume of the all particles determined and compared with the calculated mass.

The estimation of the samples, which are mounted on the sevens cantilevers, are reported in the following table with the relative quantity of electrons spins:

	X [µm]	y [µm]	Z [µm]	Mass [ng]	Nspin
Sample 1	150	100	10	1800	7.9×10^{12}
Sample 2	100	100	70	1055	4.6×10^{12}
Sample 3	72	70	70	530	2.3×10^{12}
Sample 4	40	40	80	193	8.4×10^{11}
Sample 5	30	21	19	18	7.8×10^{10}
Sample 6	11	15	20	5	2.1×10^{10}
Sample 7	5	3	6	0.1	5.9×10^{8}

Table 8.2: DPPH mounted on Nanosensors cantilevers. Seven sample of DPPH were glued on seven Nanosensors cantilevers, the mass is determined with the frequency shift and the SEM pictures and then quantity of spins calculated.

---

[8]Norland optical adhesive 65, Norland Products, P.O. Box 637, 2540 Route 130, Suite 100, Cranbury, NJ 08512, http://www.norlandprod.com.

The Q factor of the Nanosensor cantilever was found to be slightly reduced by the sample mass glued at the end of the cantilever. The measured average Q factor for all Nanosensors cantilevers is 90 K ± 20 K. In order to increase the quality factor, the cantilevers should be annealed (heated up at 800 °C in vacuum for 1-2 hours). This process cleans and reduces the defects in the mechanical single silicon beam. The sample should be then mounted. This can increase the Q factor value by 2-3, when the cantilever is annealed under UHV condition and then exposed to the air for mounting the sample.

The MRFM signal generated by sample 6 is shown in figure 8.6. The magnetic field was swept by 0.0005 Hz around a static field of 50 mT with amplitude of 6 mT. The corresponding ESR absorption line for the DPPH is also shown in figure 8.2.

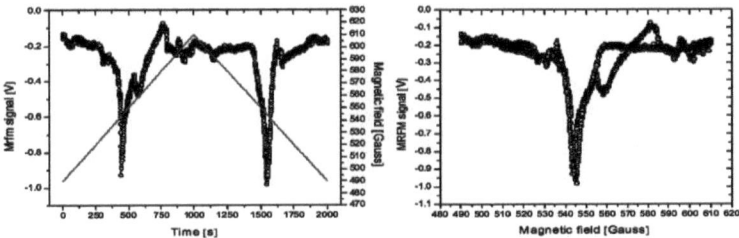

Figure 8.6: MRFM signal. These graphs show the ESR signal detected using the MRF microscope with sample 6 mounted at the end of the cantilever. The magnetic field is swept very slowly and a time constant of 1 s is chosen for the lock-in amplifier.

The figure 8.6 shows that the noise of the home built MRFM machine still has a strong amplitude at low frequency. The ESR force signal is detected from a sample with a thousand times less number of spins than the sample used in the previous standard experiment. The home built spectrometer is therefore a very interesting instrument for the analysis of substances.

The high Q factor and the long ring down time of several seconds (10 to 20 sec) necessitates a long and slow magnet sweep period of 30 min. The sweep time could be substantially decreased by adding feedback that would reduce the cantilever ring down time to the order of milliseconds [3].

8.3 Signal to noise ratio in MRFM and CW-ESR

In this section we report and compare the sensitivity experimental results made with the standard inductive ESR spectrometer and made with the MRFM machine. The results show a substantial improvement of the SNR when the force sensitivity method is used, as predicted by the theory explained in chapter 7.

All force and inductive experiments are performed at room temperature. Moreover the force experiments are performed in vacuum pressure of 10^{-6} mbar reached by a turbo molecular pump. The signal to noise ratio measured with the inductive and force methods have a linear dependency with the sample volume, which confirms the models discussed in chapter 7.

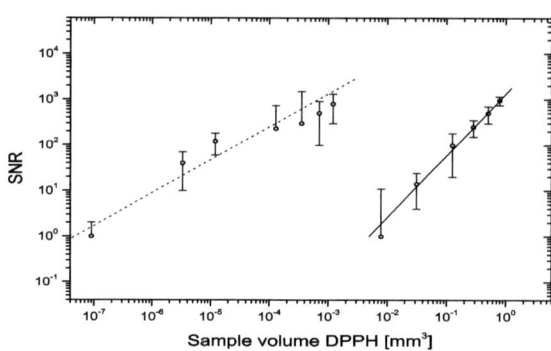

Figure 8.7: SNR of MRFM and IESR. The graph summarizes the results of two experiments performed at room temperature. The CW-ESR experiment (solid line) was conducted with 6 different samples volumes. The SNR decreases linearly when the sample volume is reduced. The dashed line shows the MRFM experiment conducted at 10^{-6} mbar with seven different samples volumes. The SNR trend is also linearly proportional with the volume.

For a constant magnetic field, the results, represented in figure 8.7, show an improvement in the SNR by a factor 100000 over the sample volumes studied. Moreover, the sensitivity difference between the inductive and force methods increases when the sample is reduced. This comparison should be performed by varying others parameters, such as the magnetic field and the temperature.

8.4 Discussion and conclusions

The experiments reported in this chapter present two different methods for detecting the electron magnetic resonance of DPPH at room temperature. A continuous wave ESR system was assembled in our laboratory using an RF microcoil. The signal from DPPH sample was measured for a decreasing series of small volume samples down to the detection limits of the system. The inductive ESR laboratory setup has a detection sensitivity of 5×10^{13} spins. This is a low sensitivity compared with recent measurements performed by other groups using a commercial ESR system [3], that report a sensitivity of 10^7 spins. The principals differences are caused by the amplitude of the static magnetic field used, by its inhomogeneity and by its time instability (electro-magnet).

In parallel, a magnetic resonance force microscopy instrument was assembled and tested in the same manner. In the first MRFM experiment [5], Rugar reported a sensitivity of 10^{11} spins, which is a sensitivity reached with our laboratory setup at room temperature. The actual limits of our CW-system are given by the low magnetic field and by the mechanical external perturbations like the turbo molecular pump.

The MRFM sensitivity can be increased of a factor of 10 by cooling down the system to the helium boiling temperature. This should stabilize the temperature and reduce the Brownian noise motion [7]. Further efforts to reduce the mechanical noise exciting the cantilever are under way, and include by using a damping table and by introducing an eddy current damping. In addition, the system must implement an ion pump in order to pump the system without any external noise. Finally, in order to increase the scanning rate a feedback loop should be implemented, so as to decrease the cantilever ring down to the order of a millisecond.

In conclusion, in this chapter we present an improvement in the SNR of MRFM over microcoil ESR by a factor of 100000. These results clearly show clearly the potential of the MRFM technique for the detecting nanogram samples and perhaps picogram masses of materials by using a room temperature microscope. The sensitivity reached with our MRFM microscope is 5.8×10^8 spins. Due of the extremely high sensitivity, the MRFM method has many applications ranging from molecular analysis to the extreme high magnetic field scanning. Moreover, MRFM pulse methods have not yet been widely studies and the technique probably has many hidden potentials.

Chapter 9
General conclusion

The thesis is divided in three main parts: the design and construction of the MRFM machine, the study of force interactions and dissipations in a vertical setup, and the detection of a large quantity of spins by the MRFM technique.

9.1 Designing and building the MRFM machine

In order to afford the ultimately limits in magnetic resonance force detection, and consequently to detect an electrons spin generating a force in atto newton range. A complete magnetic resonance force machine has been designed and assembled starting from the exciting MRFM microscope and the Janis cryostat. The assembled machine has the following characteristics: a magnetic field of 7 T, a temperature range between 1.8 K to 400 K, an ultra high vacuum (UHV) pressure (range up to $10^{-11} mbar$) and a cantilever in thermodynamical condition. Moreover, the machine is designed for working completely in UHV condition. This allows to anneal and prepare both samples and cantilevers before the measurement. The machines builded by others group do not have this important ability.

9.2 Improving the sensitivity

The force generated by a single spin experiment is in the atto Newton range. Non commercial, soft single crystalline silicon bar cantilevers with a high quality factor and minimized spring constants of 0.15 mN/m are used for detection. The extremely small spring constant makes it necessary to approach the cantilever perpendicularly near the sample surface. This setup introduces numerous phenomenological effects, which have be understood in order to reach ultimate sensitivity. The cantilever is subject in this way not only to the lateral force gradient but also to the tensile force, which are generated by electrostatic and diamagnetic interactions. Tensile force acts as the gravitational force will causes a pendulum to have an oscillation frequency proportional to the gravity. The force and force gradient has been fully calculated and approximate with a magnetic dipole. The frequency shift solution of the different eigenmodes of a cantilever subjected to a tensile force is presented. The frequency shift is calculated as a function of the diamagnetic, electrostatic forces and force gradients and compared with experimental data. Similar frequency shift graphs measured with an cantilever horizontally and vertically approaching a surface have fundamentally different force sensitivity. In the thesis we have fully analyzed the tip-sample interaction forces involves in the vertical setup.

9.3 Reduction damping losses

The cantilever tipped with hard magnetic materials is subjected to a severe damping losses, while exposed to the static magnetic field. A detection sensitivity of $10^{-18} N / \sqrt{Hz}$ in the absence on any magnetic field decreases to $10^{-16} N/\sqrt{Hz}$ at 100 mT. The principal causes of this sensitivity loss is due to the magnetic hysteresis loop oscillating with the frequency of the cantilever. Low frequency hysteresis changes do not have any effect on the Q factor. The hysteresis loop due of the inertia of the magnetic domain is inversely proportional to the frequency. We find that a cantilever oscillating at 1 kHz has more magnetic losses than the one oscillating at higher frequency. From literature we can estimate that the optimal frequency oscillation is around 5-10 kHz. With a tipped cantilever with $SmCo_5$ or $Pr_2Fe_{17}B$ the sensitivity of $10^{-18} N / \sqrt{Hz}$ is maintained constant up to 100 mT. This sensitivity should be enough for a single electron spin detection experiment.

Moreover, while the cantilever tip approaches a surface electrostatic dissipation is measured as a function of distance. The dissipation is due of the charge fluctuation of the capacitor generated from the tip and the sample surface. We calculate the electrostatic interaction assuming a spherical tip and an infinite sample surface. At a hundred nanometer distance between the cantilever tip and the surface the charge fluctuation measured is in the range of that of a single electron transistor. The charge fluctuation noise can be reduced by compensating the contact potential. Moreover, the oscillation amplitude plays a role in the energy dissipation and consequently oscillation at the thermodynamic equilibrium reduces the total dissipation. The fluctuation charges are, proportional to the oscillating amplitude of the electrostatic field.

9.4 Magnetic resonance force detection

A MRFM is not merely a form of microscopy, but in the case of nuclear spins, it is also a form of spectroscopy that can identify certain chemical elements. The extreme high sensitivity of the magnetic resonance force microscope is able to detect a single electron spin [4]. To detect individual nuclear spins and thereby achieve spatial resolution at the atomic scale, a further improvement in sensitivity by a factor of 1000 will be necessary. The potential of the room temperature magnetic resonance force microscope for molecules and biological samples is evident. In the last two chapters we build and compare a home build spectrometer and a home build magnetic resonance force microscope. A simple MRFM machine already has sensitivity improvement of more than 100000 compared to a standard ESR spectrometer.

The improvement in the signal to noise promises a huge potential of the MRFM for biological and chemical analysis.

Bibliography

[1] L. Ciobanu, D.A. Seeber, C.H. Pennington, *3D MR microscopy with resolution 3.7 µm by 3.3 µm by 3.3 µm,* Journal Magnetic resonance, 158, 178-182, 2002;

[2] K. Wütrich, The way to NMR structures of proteins, Nature Struct. Biol. 8, 923-925, 2001;

[3] A. Blank, C.R. Dunnam, P.P. Bordat, J.H. Freed, *High resolution electron spin resonance microscopy,* Journal Magnetic Resonance, 165, 116-127, 2003;

[4] D. Rugar, R. Budakian, H. J. Mamin, B. W. Chui, *Single spin detection by magnetic resonance force microscopy,* Nature, Vol 430, 15 July 2004 ;

[5] J.A. Sidles, J.L. Garbini, K.L.Bruland, D. Rugar, O. Züger, S. Hoen, C.S. Yannoni, *Magnetic resonance force microscopy,* Review of modern physics, Vol 67, No 1, January 1995;

[6] J.A. Sidles, *Folded Stern-Gerlach experiment as a mean for detecting nuclear magnetic resonance in individual nuclei,* Physic Review Letter 68, 1124-1127, 1992;

[7] U. Gysis, *Temperaturerverhalten der Elastizität und inneren Reibung mikromechanischer Resonatoren,* Thesis, Basel 2002;

[8] H. J: Mamin, R Budakian, B. W. Chui, D. Rugar, *Detection and Manipulation of statistical polarization in Small Spin Ensembles,* Phys. Rev. Letters, Vol. 91, Number 20, 14 November 2003;

[9] P. Ruff, Diplomarbeit, Universität Basel 2001;

[10] B. E. A. Saleh, *Fundamentals of Photonics,* John Wiley & Sons, Inc, New York, 1991;

[11] B. C. Stipe, H. J. Mamin, T. D. Stowe, T.W. Kenny, and D. Rugar, *Noncontact friction and force fluctuations between closely space bodies,* Phys. Rev. Letters, Vol. 87, Number 9, August 2001;

[12] U. Rabe, K. Janser, and W. Arnold, *Vibration of free and surface-coupled atomic force microscope cantilevers: Theory and experiment,* Rev. Sci. Instrum. 67 (9), September 1996;

[13] S. Rast, C. Wattinger, U. Gysin, and E. Meyer, *Dynamics of damped cantilevers,* Rev. of Sci. Instr., Volume 71, Number 7, July 2000;

[14] R. W. Clough and J. Penzien, *Dynamics of Structures*, McGraw-Hill, New York, 1975;

[15] B. C. Stipe, H. J. Mamin, T.D. Stowe, T. W. Kenny, D. Rugar, *Magnetic dissipation and fluctuation in individual nanomagnets masure by ultrasensitive cantilever magnetometry*, Physical review letters, Vol 86, Number 13, 2001;

[16] J. Mahron, R. Faichtein, D. D. Smith, *An optimal magnetic tip configuration for magnetic-resonance force microscopy of microscale buried features*, Applied physics letters, Vol. 73, Numebr 25, 1998;

[17] R. Skomski and J. M. D. Coey, *Permanent magnetism*, Institute of Physics Publishing,1999;

[18] Z. Zhang, P.C. Hammel, *Magnetic force microscopy with ferromagnetic tip mounted on the force detector*, Solid state nuclear magnetic resonance 11, 1998;

[19] J. D. Jackson, *Classical Electrodynamics*, John Wiley & Sons, 1999;

[20] L.D. Landau, E. M. Lifshitz and L. P. Pitaevskii, *Electrodynamics of continuous media*, Butterworth Heinemann, 2004;

[21] P. Grütter, Y. Liu, and LeBlanc, *Magnetic dissipation force microscopy*, Appl. Phys. Lett., Vol. 71, Number 2, 1997;

[22] Y. Liu, B. Ellman, and P. Grütter, *Theory of magnetic dissipatin imaging*, Appl. Phys. Lett., Vol. 71, Number 8, 1997;

[23] M. Krage, *Microwave sintering of ferrites*, Ceramic Bulletin, Volume 60, Nomber 11, 1981;

[24] M. E. McHenry, D. E. Laughlin, *Nano-scale materials development for future magnetic applications*, Elservier Science, Volume 48, 2000;

[25] A. Giri, K. M. Crowdary, S. A. Majetich, *AC magnetic properties of compound FeCo Nanocomposites,* Mat. Phys. Mech. 1, 2000;

[26] D. Rugar, R. Budakian, H. J. Mamin, and B. W. Chui, *Single spin detection by magnetic resonance force microscopy*, Nature, Volume 430, Number 15, July 2004;

[27] P. Lu. F. Shen, S. J. O'Shea, K. H. Lee, T. Y. Ng, Analysis of surface effects on mechanical properties of microcantilevers;

[28] G. Y. Chen,T. Thundat. E.A. Wachter and R. J. Warmack,Journal, *Absorption induced surface stress and its effects on resonance frequency of microcantilevers*, Applied Physic 77, 3618, 1995;

[29] M. Guggisberg, M. Bammerlin, CH. Loppacher, O. Pfeiffer, A. Abdurixit, V. Barwich, R. Bennewitz, A. Baratoff, E. Meyer, H. J. Güntherodt, *Separation of interactions by noncontact force microscopy*, Physical review B, Volume 61, Number 16, April 2000;

[30] S. Hudlet, M.Saint Jean, C. Guthmann, and J. Berger, *Evaluation of the capacitive force between an atomic force microscopy tip and a metallic surface*, The European physical journal B, 2, 5-10 (1998);

[31] E.R. Andrew, Nuclear magnetic resonance, Univ-Press, Cambridge, 1956;

[32] C.P. Slichter, Principles of magnetic resonance, Springer-Verlag, New York, 1990;

[33] P. Mansfield, P.G. Morris, NMR imaging in biomedicine, Academic Press, New York,1982;

[34] D.I. Hoult and R. E. Richards, *The signal to noise ratio of the nuclear magnetic resonance in a solid*, Journal of magnetic resonance 24, 71-85 (1976);

[35] A.G. Webb, *Radiofrequency microcoils in magnetic resonance*, Progress in nuclear magnetic resonance spectroscopy 31, 1-42 (1997);

[36] J.E. Stocker, T.L. Peck, A.G. Webb, M-Feng, R.L. Magin, *Nanoliter Volume, High-resoltion NMR Microspectroscopy Using a 60 mm Planar Microcoil*, IEEE Transactions on biomedical engineering, vol 44, No. 11, (1997);

[37] F. Bloch, W.W. Hansen and M. Packard, *Nuclear induction*, Physical review 69, 127 , (1946);

[38] F. Bloch, *Nuclear induction*, Physical review 70, 460-473, (1946);

[39] F. Bloch, W.W. Hansen and M. Packard, ;êxtitThe nuclear induction experiment, Physical review 70, 474-485, (1946)

[40] E.M. Purcell, H.C. Torrey and R.V. Pound, *Resonance absorption by nuclear magnetic resonance in a solid*, Physical review 69, 37, (1946);

[41] R.R. Ernst, *Sensitivity enhancement in magnetic resonance*, Advances in magnetic resonance 2, 1-135 (1966);

[42] A. Abragam, *Principles of nuclear magnetism*, Oxford University press (1999);

[43] J.D. Roberts, *ABCs of FT-NMR*, Unv. Sci. Books, Sansalito, CA,2000;

[44] R.R. Ernst and W.A. Anderson, *Sensitivity enhancement in magnetic resonance. Investigation of intermediate passage conditions*, The review of scientific Instruments 36, 1696-1706 (1965);

[45] J. D. Krauss, *Electromagnetics*, McGraw-Hill, New York, 1953;

[46] Timothy L. Peck, Richard Magin, and Paul C. Lauterbur, *Design and analysis of microcoils for NMR microscopy*, J. of magnetic resonance, Series B 108, 114-124 (1995);

[47] D. Rugar, C. S. Yannoni, and J.A. Sidles, Nature 360, 563 (1992);

[48] C.S. Yannoni, O. Zuer, D. Rugar, J. A. Sidles, *Force detection and imaging in magnetic resonance*, Encyclopedia of nuclear magnetic resonance, New York, Wiley, 1996;

[49] A.Suter and all., *Probe-sample Coupling in the magnetic resonance force microscopy*, Journal of magnetic resonance 154, 216-227 (2002);

[50] Todd G. Ruskell, Markus Löhndorf, John Moreland, *Field mapping with the magnetic resonance force microscope*, Journal of applied physics, Vol 86, Number 1, 1999;

[51] O. Züger, D. Rugar, J. Applied Physics 75, 6211, 1994;

[52] K.J. Bruland, J.L. Garbini, W.M. Dougherty, J.A. Sidles, *Optimal control of force microscope cantilevers. Magnetic coupling implementation*, J. Appl. Phys. 80, 1996;

Acknowledgements

First at all, I thank Prof. Dr. E. Meyer, head of the MRFM group and responsible of the module IX of the NCCR project. I thank him for give me the opportunity to work in the new field of magnetic resonance force microscopy and for the continues interest and support. I like to thank Prof. Dr. H. J. Güntherodt, head of the scanning probe group in Basel and head of the National Center of Competence (NCCR), for the continuous interest in the project and for the support he provided me in many ways. I thank Dr. S. Rast, the responsible of the MRFM machine group, and Dr. U. Gysin for teaching me tricks of magnetic resonance force microscope and general physics. I thank Prof. Dr. H. J. Hug and Dr. Ch. Gerber for their useful help in the design of the MRFM machine. I thank Prof. Dr. Baratov for discussion about the interpretation of the results. I thank M. Sasha the technician of the MFM Dr H. J. Hug group and the actual chef of the mechanical shop, for the great experience help. I thank J.P. Ramseyer our technician for the great help and for learn me "la bonne cuisine franaise". I thank R. Hamid, a PhD student, and P. Ruff, our actual technician, for the exchange during my thesis. I would like to thank all past and present co-workers in the physic group, whose I share the PhD. time and I have discuss of the project: Dr. E. Gnecco, Dr. L. Nony, Dr. A. Wetzel. I would like to thank the members of our electronic shop, the head H.R. Hidber, A. Tonin and R. Maffiolini, for their friendly support. I thank M. Steinacher head of the electronic group for the great help and fine work. I would like to thank H. Breitenstein and J. Silvester for their great mechanical help. Especially, W. Roth for the evening help and for the important mechanical machine explanation and for providing liquid nitrogen and helium even late in the evening. I like to thank my friends Dr. B. Gimi, Dr. B. Roman and Prof. Dr R. Magin for providing me great English corrections. I thank my parents and my brother for their continuous support during the thesis. Finally a special thank to Stefania, my girlfriend, for help me in the corrections of the thesis and for sharing the PhD time. THANKS

Chapter A

Annexes

A.1 Beam Flexure: Including axial force effect

The axial force may have a very significant effect on the vibration of the cantilever, resulting in a shape and frequency modification. The motion equation including the effect of an axial **N** force which is uniform along the cantilever and is not varying with the time becomes:

$$EI\frac{\partial^4 v}{\partial x^4} + N\frac{\partial^2 v}{\partial x^2} + \bar{m}\frac{\partial^2 v}{\partial t^2} = 0 \qquad (A.1)$$

By assuming harmonic motion solution and by the separation of variables the equation 11.1 reduces to two independent equations:

$$\ddot{Y}(t) + \omega^2 Y(t) = 0 \qquad (A.2)$$

$$EI\phi^{iv}(x) + N\phi(x) - \bar{m}\omega^2\phi(x) = 0 \qquad (A.3)$$

In which E is the modulus of elasticity, I is the moment are, N is the normal force, m the linear mass and ω is the frequency.

The first equation A.2 is the time variation equation showing that a constant axial force does not affect the simple harmonic character of the free vibrations. The second equation A.3 leads to frequency and mode-shape expression for a free beam cantilever, which N is a constant parameter. The second equation divided by EI can be written:

$$\phi^{iv}(x) + g^2\phi(x) - a^4\phi(x) = 0 \qquad (A.4)$$

in which

$$a^4 = \frac{\bar{m}\omega^2}{EI} \qquad g^2 = \frac{N}{EI} \qquad (A.5)$$

The solution of equation can be obtained in the standard way by assuming a solution in the form:

$$\left(s^4 + g^2 s^2 - a^4\right) C e^{sx} = 0 \qquad (A.6)$$

in which
$$s = \pm i\delta, \quad \pm\varepsilon \qquad (A.7)$$
and
$$\varepsilon = \sqrt{\left(a^2 + \frac{g^4}{4}\right)^{\frac{1}{2}} - \frac{g^2}{2}} \qquad \delta = \sqrt{\left(a^2 + \frac{g^4}{4}\right)^{\frac{1}{2}} + \frac{g^2}{2}} \qquad (A.8)$$

The final shape expression of the exponential equivalent function is a sum of trigonometric and hyperbolic expression:

$$\phi(x) = D_1 \sin(\delta x) + D_2 \cos(\delta x) + D_3 \sinh(\varepsilon x) + D_4 \cosh(\varepsilon x) \qquad (A.9)$$

The coefficients D_1, D_2, D_3, D_4 can be evaluated by consideration of the boundary conditions which are described as:

$$\begin{aligned}\phi(x)|_{x=0} &= 0 \\ \frac{\partial \phi(x)}{\partial x}\Big|_{x=0} &= 0 \\ \frac{\partial^3 \phi(x)}{\partial x^3} - g^2 \frac{\partial \phi(x)}{\partial x}\Big|_{x=l} &= 0 \\ \frac{\partial^2 \phi(x)}{\partial x^2}\Big|_{x=l} &= 0\end{aligned} \qquad (A.10)$$

Substituting the boundary into equation , we obtain the corresponding frequency equation as

$$\frac{\varepsilon \sin(\varepsilon l) + \delta \sinh(\delta l)}{\varepsilon^2 \cos(\varepsilon l) + \delta^2 \cosh(\delta l)} = -\frac{\delta^2 \cos(\delta l) + \varepsilon^2 \cosh(\varepsilon l)}{\varepsilon \delta^2 \sin(\delta l) - \delta \varepsilon^2 \sinh(\varepsilon l)} \qquad (A.11)$$

By letting N be zero, the above equation reduces to the frequency of the cantilever without axial force

$$\cos(al)\cosh(al) = -1 \qquad (A.12)$$

With an axial force N, the related eigenvalues can be obtained from . The frequency of the beam will be

$$\omega_n = \frac{(k_n l)^2}{l^2}\sqrt{\frac{EI}{\bar{m}}} \qquad (A.13)$$

In our problem, the values of $k_n l$ depend of the axial force N. It can be calculated that when the axial force increase the natural frequency increase, and for a compressive force the frequency of the transverse frequency decrease.

A.2 Design of RF coax cable

The coax cable in the gradient temperature region has been home built. The characteristic of the transmission loss power pro meter of the coax cable built is presented in the graph A.1. The Coax 1, Coax 2 and Coax 3 represent the three RF cables mounted in the high gradient temperature region. The Coax 1 and 2 are not gold plaited, because the coating could improve the heating flow. The Coax 3 is at 60 cm over the cone, consequently has been gold plaited for increasing the RF transmission.

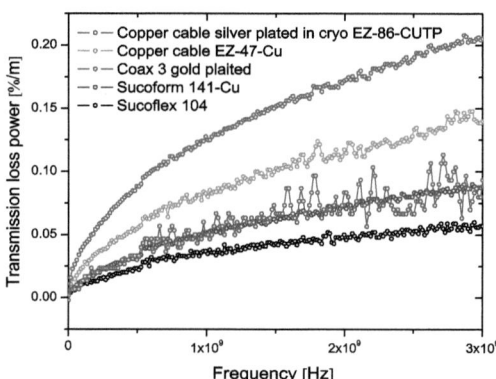

Figure A.1: The graph represent the power transmission of commercials cables, and of the home built cable

The coax 3 has a transmission power loss very close to the best cable of Huber & Suhner available in the market (Sucoflex 104). The design of the cable is very simple, two sma RF connector of Huber & Suhner as been used and connected to two pipes of stainless steel 304. The external diameters of the internal conductor is ϕ_{in}= 3 mm and the internal diameter of the external of ϕ_{ext}= 7 mm. A characteristic impedance of 50 Ω meets with two concentrically conductor with the following relation:

$$Z = \frac{1}{2\pi}\sqrt{\frac{\mu_0}{\varepsilon_0}} \ln\left(\frac{r_{ext}}{r_{in}}\right) = \frac{1}{2\pi}\sqrt{\frac{12.56 x 10^{-7}}{8.85 x 10^{-12}}} \ln\left(\frac{3.5}{1.5}\right) \cong 50\Omega \qquad (A.14)$$

A.3 Cryogenic System

We report the whole scheme of the entire magnetic resonance force microscope. The complete design has more than hundred of sketch.

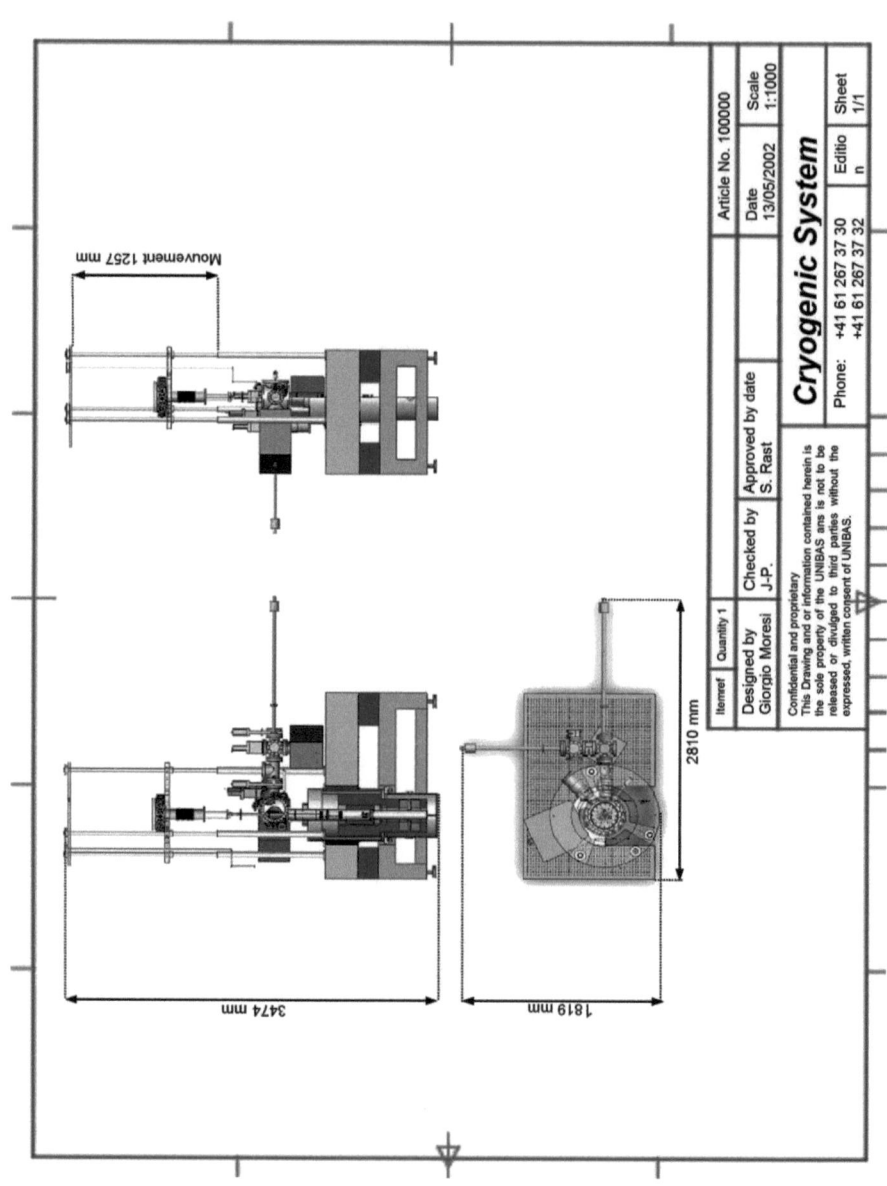

Curriculum vitae

28 Mai 1975	Born in Sorengo (Switzerland, TI), son of Iza Moresi geb. Regazzoni and Aldo Moresi
1981-1986	Elementary school in Bellinzona, Switzerland.
1986-1990	Middle school in Bellinzona, Switzerland.
1990-1994	Maturity type C, High school, Bellinzona, Switzerland.
1994-1999	Studies of Microtechnical Engineering, EPFL, Lausanne, Switzerland. Specialization: Integrated Products.
Feb 1999	MSc. in Microtechnical Engineering, Integrated Products, Institute of Applied Optics (IOA), Swiss Federal Institute of Technology Lausanne and CSEM Zürich, Switzerland, Prof. Dr. R. Salathé and Prof. Dr. P. Seitz
1999-2000	Research assistant at the Institute of Microsystems, EPFL, Lausanne, Switzerland.
2000-2002	Researcher at the Bioengineering department, UIC, Chicago, USA.
2002-2005	PhD. Phil. Nat., thesis in Physics, Condensed Matter Division, University of Basel, Switzerland, Prof. Dr. E. Meyer, Prof. Dr. Hans-Joachim Güntherodt and Prof. Dr. H. Hug

i want morebooks!

Buy your books fast and straightforward online - at one of world's fastest growing online book stores! Environmentally sound due to Print-on-Demand technologies.

Buy your books online at
www.get-morebooks.com

Kaufen Sie Ihre Bücher schnell und unkompliziert online – auf einer der am schnellsten wachsenden Buchhandelsplattformen weltweit! Dank Print-On-Demand umwelt- und ressourcenschonend produziert.

Bücher schneller online kaufen
www.morebooks.de

VDM Verlagsservicegesellschaft mbH
Heinrich-Böcking-Str. 6-8   Telefon: +49 681 3720 174   info@vdm-vsg.de
D - 66121 Saarbrücken   Telefax: +49 681 3720 1749 www.vdm-vsg.de

Printed by Books on Demand GmbH, Norderstedt / Germany